ISRAEL, the
UNITED STATES,
and the War Against
HAMAS
July–August 2014

To my granddaughters
Avigail and Ella

ISRAEL, *the* UNITED STATES, *and the War Against* HAMAS

July–August 2014

The "Special Relationship" under Scrutiny

Zaki Shalom

sussex
A C A D E M I C
P R E S S
Brighton • Chicago • Toronto

2 4 6 8 10 9 7 5 3 1

First published in hardcover 2019, reprinted in paperback 2020, by
SUSSEX ACADEMIC PRESS
PO Box 139
Eastbourne BN24 9BP

Distributed in North America by
SUSSEX ACADEMIC PRESS
Independent Publishers Group
814 N. Franklin Street, Chicago, IL 60610

British Library Cataloguing in Publication Data
A CIP catalogue record for this book is available from the British Library.

Library of Congress Cataloging-in-Publication Data
Names: Shalom, Zaki, author.
Title: Israel, the United States, and the war against Hamas, July-August
2014 : the "special relationship" under scrutiny / Zaki Shalom.
Description: Brighton ; Chicago ; Toronto : Sussex Academic Press,
    [2019] | Includes bibliographical references and index.
Identifiers: LCCN 2018057601 | ISBN 9781845199890 (hbk : alk. paper)
                    | ISBN 9781789760491 (pbk : alk. paper)
Subjects: LCSH: Gaza War, 2014—Foreign public opinion, American. |
    Public opinion—United States. | Israel—Foreign relations—United
    States. | United States—Foreign relations—Israel.
Classification: LCC DS119.768 S53 2019 | DDC 956.9405/5—dc23
LC record available at https://lccn.loc.gov/2018057601

Typeset and designed by Sussex Academic Press, Brighton & Eastbourne.
Printed by TJ International, Padstow, Cornwall.

Contents

Acknowledgments

The birth of this research began in the days that followed Operation Protective Edge, as the war itself brought to light many questions for Israel: How is it possible that the formidable Israel Defense Forces, with its overwhelming capabilities and advanced technology, could not defeat a relatively small terror organization – Hamas – quickly and soundly within a few days of fighting? Why was Israel unable (or unwilling) to fully utilize its military superiority throughout the conflict; can a regular army in a democratic state defeat a terror organization? This book, however, focuses on another crucial question that was raised during the war and after it: What was the role of the United States during Operation Protective Edge? More accurately – did it really act as a state that keeps "special relations" with Israel?

First and foremost, I would like to thank the director of the Institute for National Security Studies (INSS), Maj. Gen. (ret.) Amos Yadlin. I have learned much from his analysis of strategic issues, as well as from other staff members at the Institute for National Security Studies; within the many meetings, sessions, conferences held at INSS and elsewhere.

I am very grateful for the staff at the INSS information center, headed by Yoel Kozak, for their endless efforts to supply me with the information and knowledge required for this research. I would also like to thank my roommate and close friend at the institute, Dr. Kobi Michael, for his insights on the issues that surround this topic. Additionally, I would like to give thanks to my old friend at INSS, Moshe Grundman, for his advice and efforts in the composition of the cover page.

Finally, I wish to extend my thanks to the many research assistants at the INSS, who greatly helped me in writing this book. I would like in particular to mention Bridget Mudd, who significantly helped me in the gathering of information and details related to my subject of research. Special thanks are also extended to my highly talented and motivated assistant in recent months – Jacob Aaron Collier. His depth of knowledge and his ability to phrase complicated issues in a very simple and clear manner were crucial contributions in the process that led to the publication of this book.

The cover illustration shows U.S. President Obama and Prime Minister Netanyahu, photographed with military personnel, Ben-

Gurion Airport, Lod, Israel. The Iron Dome anti-missile system is in the background. Photo by Moshe Milner, courtesy of The Government Press Office Photography Department, February 30, 2013.

Introduction

On July 8, 2014, the Israeli Defense Forces (IDF) launched Operation "Protective Edge" after a series of escalating violent incidents between Israel and the Palestinians.

Over the summer of 2014, violence between Israel and the Palestinians escalated, reaching its climax with the kidnapping and murder of three Israeli teenagers. At the same time, Hamas greatly increased rocket and mortar fire targeting Israeli civilians, particularly in the southern part of the state – the Negev. This clearly reflected the weakening of Israel's deterrence vis-à-vis Hamas. At this period of time, Hamas most likely estimated that Israel would tolerate a high level of Palestinian provocations in order to avoid engagement in an all-out confrontation. Israel clearly had no interest in any escalation with Hamas at that time, as it had nothing to gain from a major military confrontation. As long as there was no intensive violence across the border, Israel accepted the current status quo.

Hamas clearly maintains an effective control over the life of the people in Gaza since it came to power in 2007. Even now, when Israel is totally out of Gaza, following the implementation of the disengagement plan of 2005, and does not have any military or civilian presence there, it is often blamed by the international community for the poor economic situation that prevails there: "The disengagement plan," stated Israel Katz, Minister of Transportation, "has created two major problems for Israel. The first one is the rise of Hamas to power in Gaza, which made Gaza a base of terror against Israel. The second problem is the fact that notwithstanding its withdrawal from Gaza, Israel is still held as being responsible for the civilian suffering of almost two million people in Gaza. Egypt, which has a border with Gaza, does not want to undertake responsibility for Gaza, and it pushes her to Israel's wide arms. Thus Israel is seen as dominantly responsible for Gaza as if it is still controlling the territory as an occupying power."[1]

One can only imagine the amount of criticism that will be launched against Israel once it becomes known that Israel possesses a de jure and de facto control over the Gaza Strip. Furthermore, Israel's greatest fear is a state of anarchy in the occupied territories where there is no entity that undertakes responsibility for what takes place. If Israel has to choose between a rule of a radical fanatic Islamic regime over Gaza and a state of anarchy in that territory, it would certainly prefer the first option, simply because it can establish deterrence against a radical regime, while in a situation of anarchy it is powerless.

The present Israeli government operates under constant fear that an international coalition led by the United States and the European Union would exert heavy pressures on Israel to withdraw from the West Bank without gaining in return a comprehensive lasting peace with the Palestinians. It should be recalled that every American administration since the Six Day War adopted a formula calling upon Israel to withdraw from all of the territories it had occupied during the war of 1967, allowing only "minor modifications" in the borders needed to ensure Israel's vital security interests. President Bush was the first to deviate from this position. On April 14, 2004, President Bush sent a letter to Prime Minster Sharon laying out a new settlement framework. According to Bush, given the construction of large settlements in the occupied territories it would be unrealistic to expect Israel's permanent borders to return to the 1949 armistice lines. The implication was clear: "There would be no return to 1967 and Israel could keep the major settlement blocs. Various groups in Israel interpreted this position as the administration's indirect consent to continuing the settlement enterprise within the existing Jewish settlement blocs in Judea and Samaria."[2]

President Obama chose to ignore the Bush administration's commitment to Israel, as his administration seemed to be more determined than the previous one to bring about a comprehensive Israeli–Palestinian peace agreement. It believed that an agreement was "a vital national security interest of the United States." According to their position, the Israeli–Palestinian conflict was the main source of problems for the US in the Middle East. A resolution would improve US relations with Arab and Muslim nations, and thereby also improve the regional standing of the United States. The president expressed his position openly as early as the 2008 election campaign. In an interview with Tom Brokaw on *Meet the Press* on July 27, 2008, he argued that the United States had to adopt an "overarching strategy" in the Middle East, recognizing that all of the region's problems were interconnected. If the Israeli–Palestinian conflict was resolved, the Arab states would strengthen ties with the United States, enabling the Americans to then exert more power in solving other regional crises, such as those in Iraq, Afghanistan, and Iran.[3]

According to Obama, "If we've gotten an Israeli–Palestinian peace deal, maybe at the same time peeling Syria out of the Iranian orbit, that makes it easier to isolate Iran." Several of Obama's top advisors agreed with this position, creating an echo chamber within the White House. National Security Advisor Jim Jones noted, "I'm of the belief that had

God appeared in front of President Obama in 2009 and told him he could choose one thing on the face of the planet, and one thing only, to be done, to make the world a better place and give people more hope and opportunity for the future, I would venture that it would have something to do with finding the two-state solution to the Middle East."[4]

Only recently Russian Foreign Minister Sergey Lavrov revealed that President Obama in fact wished to impose his peace plan, which was considered by Israel as extremely harmful to Israel's national interests, upon the parties; namely upon Israel: At the end of 2016, Lavrov said, 'that the Obama Administration wished to impose artificial parameters with regard to the Israeli–Palestinian settlement.' Russia did not support this initiative of the United States, considering that it would inevitably lead to a situation 'on the ground' significantly aggravated and without real impact." Lavrov explained, "It would be counterproductive." "And during   a long telephone conversation at the end of December with the US   secretary of state, we clarified the Russian position, which was also taken into account at the request of [Prime Minister of Israel] Benjamin Netanyahu," he noted.[5]

In general, most Israeli governments since 1967 have viewed the so-called "peace process" with suspicion. They rightly estimated that such a process would necessarily lead the international community to exert heavy pressure on Israel to withdraw from the occupied territories to the June 5, 1967 borders. This position has been prevalent within the international community, and in particular the United States and Europe, since the Six Day War. The Israeli perception is that Israel would have to make real concrete concessions – to withdraw from territories, to expel Israeli residents from their homes – while the Palestinians would be asked only to make appeasing moderate statements. Furthermore, Israel's acts would be irreversible. Israel would not be able to restore its control over the territories from which it had withdrawn. The Palestinians, on the other hand, could remove themselves from any statement they had pledged commitment to. They have done so many times in the past, and there was no reason to believe that they would not act the same way in the future.

From the point of view of the government in Israel, the rule of Hamas over Gaza has some positive aspects for Israel, despite its extreme hostility towards the Jewish state. Due to the rule of Hamas over Gaza and the separation between Gaza and the West Bank, Israel can claim, and indeed it does, that there is already one Palestinian state in Gaza. Therefore, the talks with the Palestinian entity on the two-state solution would bring about the establishment of two Palestinian states – one in Gaza and one in the West Bank. Israel can rightly argue, that it cannot accept such a development that poses grave strategic dangers. Here we see that the existence of the Hamas regime in Gaza helps the present government in Israel to struggle against the establishment of a Palestinian state in the West Bank. The present right-wing government in Israel perceives such a development to be a major danger to Israel's strategic interests. Consequently, it would certainly prefer the continuation of the present status quo.

Present Israeli leadership under Benjamin Netanyahu seems to estimate that time is working in favor of Israel's strategic interests. In the first place, it believes that the balance of power between Israel and its neighboring enemies are shifting in favor of Israel. Throughout the years since the establishment of the State of Israel, the dominant assessment was that time is not working in favor of Israel. As time passes, the number of Arabs in countries surrounding Israel would grow more than the growth of Jews in Israel. Furthermore, dictators and monarchs, who are able to force their citizens to obey any policy they wish to implement, rule over much of the Arab world. Israel, on the other hand, as a democratic state, inherently suffers from greater division. Given this assessment, it would not be illogical to suppose that all of these components would necessarily ensure that the Arab world would enjoy superiority over Israel.

However, over time many of the advantages thought to be held by the Arabs against Israel lost relevance, particularly in the context of modern warfare, where quality often overcomes quantity. Despite its small size, Israel punches far above its weight, having established one of the most technologically advanced armed forces in the world. It currently spends 4.5 percent its GDP on research and development, with 30 percent of that amount going towards products of a military nature, and by averaging $6.5 billion worth in arms sales every year, Israel is also one of the top weapons exporters in the world.[6] While the Arabs have always enjoyed significant numerical advantages over their Jewish neighbors, the Arab world is far from being a unified force, and Israel's rise as a technological superpower has put it head and shoulders above its Arab neighbors as the dominant military power in the Middle East. Much of this can be contributed to the creativity and innovation that Israel has so heavily relied on for its survival and success over the years, especially given its lack of natural resources. Another reason why Israel has grown so adept at defending itself is that it has now 70 years of experience fighting for its very survival. On the diplomatic front, Israel's peace agreements with Egypt and Jordan have remained intact, and the division and relative weakness of Israel's other neighbors have rendered traditional enemies (such as Syria) unable and/or unwilling to confront Israel directly.

As time passes Israel's hold over the West Bank is rapidly growing, thus making the fulfillment of the aspiration of right-wing parties in

Israel to make the West Bank areas (Judea and Samaria in their language) an integral and eternal part of the State of Israel a concrete option. The number of settlements and Israeli families who wish to live in the West Bank is expanding as the costs of living inside Israel are mounting. Many Israeli youngsters wish to live in the West Bank not because of ideological or religious reasons, but because this enables them to highly improve their standard of living in various respects. Living in the West Bank enables them to live nearby the center of Israel, while the cost of living is much lower. They would be able to grant their children a high standard of education in a countryside environment. Nowhere in Israel can they get such a tempting social-economic "basket".[7]

In 2015 there were almost half a million settlers living in the West Bank, including Jerusalem. The supporters of the settlement project believe that within a few years there will be close to one million Jews living in the West Bank. When this happens, they believe, the two-state solution will no longer be a relevant option. The rise of right-wing parties in Europe and the prospect that president Trump would be reelected encourage their belief that the settlement project in the West Bank will continue apace without interruption, and even expand. The days when the international community would condemn Israel for each house being built in the West Bank, they believe, are over. The United States and Europe do not give Israel "free hand" in its settlement activities in the West Bank, but they are much more flexible than in the past, and tend to "close eyes" about the development of the settlement project, as long as it is being carried out in a "normal dimension".

The settlement project first began in 1968. At that period of time there were only five settlements in the West Bank with about 300 settlers. Those settlements were established by the Labor Party, which considers itself to be a center-left wing party. The big change occurred following the rise to power of the right-wing party, the Likud, headed by Prime Minister Menachem Begin in 1977. By that time there were 38 settlements in the West Bank, in which there were 1,900 settlers. Ten years later, the number of settlers grew to 50,000 living in 100 settlements. In 1997, there were already over 150,000 settlers in the West Bank. Today, as mentioned, the number of Jewish settlers living in the West Bank, including East Jerusalem, is almost 600,000.[8]

Another reason that contributed to the growing numbers of Jews who wish to settle in the West Bank is the improvement of the security situation there. The fear of Israelis that living in the West Bank would highly jeopardize their security has been gradually eroding in recent years. For a long period of time, after the Six Day War, the situation in the occupied territories had been relatively calm. The First Intifada, which began in 1987, and in particular – the Second Intifada, which began in 2000, shattered this sense of relative tranquility, and posed a grave challenge to Israel's security.[9]

Throughout the years since the end of the Six Day War, there were periods in which there were many events of intensive Palestinian terror not only in Gaza and the West Bank but also in the main cities of Israel. The worst and most dangerous form of terror against Israel was the one that was carried out by suicide terrorists. Many of them came to the centers of Tel Aviv, Haifa and Jerusalem. They masked themselves as regular Israelis and they exploded themselves while killing many civilians. During the Second Intifada (September 2000–December 2005) there were 147 suicide acts of terror in which 525 Israelis, most of them civilians, were killed; 3,350 Israeli were wounded. The term "suicide" does not really reflect the character of this act of terror. In the western thinking, suicide is an act that reflects a state of despair by a person, who wishes to die rather than continue to suffer. The Palestinians who carry out a terror act of suicide believe that they are doing a sacred mission by sacrificing their life to G-d. Therefore, they believe that they will be generously rewarded by G-d after their death. The most popular targets for those acts were buses or bus stations. Other targets were restaurants, cafés, shops, and malls. Most of these acts were carried out in Jerusalem and then in nearby Tel Aviv. The numbers: 425 of the people killed during this period of time were civilians; 73 of those killed were security personnel; 56 percent of those killed were men and 44 percent were women. Most of the suicide terrorists were educated people; men carried out most of the attacks, however, women carried out eight of these attacks as well.[10]

Israel's defense authorities found it difficult to cope with this form of terror, which had grave social-economic and moral effects on Israeli society. People were afraid to ride on buses, to sit in cafés, or to buy food in the markets. The Israeli economy suffered great losses during those years. Foreign investors and tourists hesitated before coming to Israel. Obviously, the attacks also weakened Israel's political position within the international community. The widespread belief in the "unlimited" abilities of the Israel Defense Forces was severely damaged.

Worst of all was the fact that for a long period of time, in particular during the Second Intifada, the security establishment in Israel had no idea who should exactly be defined as Israel's enemy. Israel had signed a treaty with the Palestinian entity known as "the Oslo Agreement". In this agreement the parties undertook to refrain from violent acts against the other. However, intelligence reports had clearly indicated that the Palestinian entity and its leader Yasser Arafat were intensively involved in the terror acts against Israel. During Operation "Defensive Shield," Israel found many documents that proved that the Palestinian entity and

Arafat personally were involved in the terror acts against Israel at three levels:

1　The ideological level: an intensive effort to convince Palestinians to "volunteer" for suicide acts against Israel, telling them that they would be carrying out a sacred mission for Islam, and they would be generously rewarded.
2　Financial support: the documents clearly showed that the Palestinian entity was engaged in delivering money needed for the carrying out of terror acts against Israel.
3　Training for terror acts: the documents proved that the Palestinian entity was heavily engaged in training terrorists, giving guidance and intelligence information needed for the implementation of terror acts.[11]

Arafat denied any involvement in terror and tried to blame extremist Palestinian organizations for these terror acts. Those organizations, he claimed, were his enemies as well as Israel's enemy.[12] There were many in Israel and in the international community who believed him, and considered him to be a true partner for Israeli–Palestinian conciliation. They claimed that the Palestinian entity was cooperating with Israel's security authorities and even arresting people suspicious for intent to carry out acts of terror. Shimon Peres, who initiated the Oslo agreements, tried to encourage the Israeli people following acts of terror carried out by Palestinians: Most of the Israelis," he stated, following a horrible act of terror on a bus in Jerusalem, "know that peace has price in human life".[13]

Although the Palestinian entity was cooperating with Israel at a certain level against terror, this did not prevent Arafat from encouraging terror against Israel. He knew very well how to play this double game. Even while conducting peace talks with Israel, Arafat was directing terror acts against Israel. The Palestinian entity did arrest people who were engaged in terror acts against Israel. However, they used the trick of "revolving doors"; namely, shortly after being arrested they were released from prison.[14]

The other issue that created deep controversy within Israel during the Second Intifada related to the question of how this form of terror could be stopped. Many in Israel believed that this terror could not be eliminated by military means since it only reflected the legitimate aspirations of the Palestinians to have a state of their own. One can fight and even defeat, they claimed, a terror organization. But no one can eliminate a national movement that wishes to fulfill the legitimate aspirations of its people. They believed that once Israel would be ready to withdraw from the occupied territories and grant the Palestinians a state of their own, the Palestinians would no longer be motivated to fight against Israel and the terror acts against Israel would end. They reminded the Israeli people that many states that gained their independence during the twentieth century have achieved their goal through acts of terror in conjunction with diplomatic efforts. [15]

Even the Jewish community's struggle for independence, they claimed, included intensive acts of terror against Britain – intended to force her to end its control over Eretz Israel during the thirties and forties of the twentieth century. Israel, they claimed, cannot seriously try to delegitimize the Palestinian struggle to realize their national aspirations given that it had taken the same measures in order to gain its own independence. A leading Israeli general, Amram Mitzna, frequently reflected the view that the Palestinians were struggling for their national independence. These aspirations cannot be destroyed by military force. Israel must negotiate with them a political solution that will fulfill their dream, and thereby ensure Israel's security. "Our warfare against the Palestinians," he claimed, "exhausts all our resources. More and more Israelis understand now that there is no military solution to this conflict. All those who think that if only we give the army free hand, we will win the struggle, are totally wrong. These are popular slogans, but they are just illusion."[16]

Moreover, security officials and security experts claimed that those who carry out the terror acts, especially the suicide terrorists, are highly motivated people. They are aware of the fact that they will die one way or another once the act of terror has been carried out. There is no way, they argue, that you can deter someone who is ready, even willing, to die for the cause of his or her struggle. Once they are killed, they become martyrs and a symbol for many young Palestinians who wish to gain the same admiration as those who were "murdered by the Zionist conquerors." Consequently, major General Dan Shomron, the Chief of staff during the First Intifada made the following controversial statement:

> The violent struggle of the Palestinian people can be stopped only by undertaking extremely brutal measures such as a transfer of population to other states or by ways of genocide. None of these measures are acceptable in Israel. The IDF cannot eliminate a national struggle carried out by the Palestinian people. It, at best, can lower the profile of the flames.[17]

Finally, those who supported the political track, as opposed to the military track, in the struggle against Palestinian terror operations, claimed that a military struggle against terror would necessarily entail harm to innocent people. This will only inflame the hatred towards Israel and enhance the terror against Israelis. Thus, instead of lowering the level of antagonism, violent acts of retaliation carried out by Palestinians will only rise up. Furthermore, the killing of innocent people, in particular women and children, will necessarily damage Israel's image in world public opinion. There is no way you can explain, let alone justify, the killing of a young child in Gaza. The bottom line is clear: there is no alternative to a political agreement with the Palestinians. Some leading Israeli Air Force (IAF) pilots claimed that since the direct targeting of leading Palestinian figures would necessarily involve the killing of innocent people, it was therefore both an illegal and unmoral act.[18] This was interpreted as a call to young pilots in the Israeli Air Force to refuse fulfilling an order to carry out direct targeting against Palestinian leaders or officers, which would necessarily entail, even if unintended, the killing of innocent people.[19]

Leading figures in the Israeli security establishment refused to accept these arguments that would eventually lead to a Palestinian victory over Israel. In the first place, they argued, the national movements that fought for their independence during the forties and fifties of the twentieth century carried out their struggle against colonial powers that had occupied their land, though they had no historic or religious attachment to it. Israel is not a colonial power. The Jewish people returned to their ancient homeland from which they had been exiled two thousand years ago. The main cities in the West Bank, like Hebron, Bethlehem, Nablus, and Jerusalem, are all ancient Jewish towns that had been under Jewish rule for long periods of time. There can be no comparison between the rule of Israel over those cities and the rule of a colonial power like Britain in India or Egypt. An Israeli scholar said the following in this regard:

It is clear why the Palestinians describe their conflict with Israel as a conflict of a nation against a colonial power who occupied their land. This gives them a justification to undertake any measure they wish in their struggle against the colonial power. It should be noted that the Palestinians used their rhetoric against Israel as a colonial power before the outbreak of the Six Day War and the occupation of territories. Our conflict with the Palestinians, however, is basically a national conflict where two national movements think the territory of "Eretz Israel" belongs to them. No colonial power has struggled for 50 years to keep their control over the territory. Those in Israel who support the Palestinian view of the conflict as a colonial conflict only help to convince the Israelis that a withdrawal from the territories will not bring an end to the conflict, but rather will bring the struggle closer to Tel Aviv.[20]

Moreover, many Israelis believed that the Palestinian national movement does not aspire just to have a state for the Palestinian people, within the June 5, 1967 lines, which would live side by side with the only Jewish state in the world – the State of Israel. The Palestinian leadership wishes to annihilate the State of Israel and replace it with a Palestinian state. That is why the diplomatic and political course cannot solve the Israeli–Palestinian conflict. Israel is doomed to fight against the Palestinians for many years to come until they become resigned to the legitimacy of a Jewish state in this region. The opposition of the Palestinian leaders to recognize Israel as the state of the Jewish people, Netanyahu and other right-wing Israeli officials stress, clearly proves that the aim of the Palestinian entity is not to establish an independent state within the June 5, 1967 lines side by side with the Jewish state of Israel.[21]

Many within the Israeli security establishment further claim that Israel has the ability to defeat the Palestinian terror. The ethos of the Israeli army is built upon the heritage of Prime Minister Ben- Gurion. It is based on the principle that Israeli security forces must be capable of coping with any threat posed against Israel. Eventually, they were right. Following a horrible terror event in Netanya at the Park Hotel (the Passover massacre) on March 27, 2002, in which 30 civilians were killed and over 140 were injured, Israel initiated an overall attack against the terror infrastructure in the West Bank in what became known as Operation Defensive Shield. Prime Minister Netanyahu stated the following in this regard:

> There were many who thought we cannot defeat terror. That there is no military solution to terror. Operation Defensive Shield proved that we can defeat terror, that there is a military solution to terror. The call for a political solution forces us only to make concessions to the Palestinians. These concessions only demonstrate our weakness and severely decreases our deterrence image in front of our enemies.[22]

Following this operation, the security situation in the West Bank improved. Still, in the eyes of many Israelis it is much more dangerous to live in the West Bank than in cities inside Israel. Nevertheless, the population in the West Bank is growing rapidly, thus diminishing the chances of an Israeli withdrawal from the area.

Since Operation Defensive Shield the level of Palestinian terror against Israel has gradually decreased, and though there are occasional events of terror against Israelis, these acts do not significantly affect the daily life in Israel. They certainly do not pose a strategic threat to Israel. Indeed, many of these incidents occur in the West Bank. As noted, this does not affect the willingness of many Israelis to build their home in the West Bank.

The present Israeli leadership under Prime Minister Netanyahu estimates, with much justification, that any change in the present status quo in the occupied territories would be detrimental to Israel. It would force her to withdraw from the present lines in the West Bank. There seems to be a broad consensus in Israel that the chances of a real peace between Israel and the Palestinians are extremely low under the present circumstances, even if a left-wing party would come to power in Israel. The director of the Institute for National Security Studies, General Amos Yadlin, who is far from being identified with right-wing thinking, stated recently that the chances for an Israeli–Palestinian peace agreement are very low.[23]

Under these circumstances, an Israeli withdrawal will almost certainly lead to a situation in which the control over the territories from which Israel withdrew would be in the hands of radical Islamic groups supported by extremist states such as Iran and Syria. That's what happened in Lebanon after Israel withdrew from southern Lebanon in May 2000. The Israeli government at that time, headed by Prime Minister Ehud Barak, came to the conclusion that the continued presence of the IDF in Lebanon did not enhance Israel's security. In fact, it weakened Israel's position. Implicitly, the government admitted that Israel's security authorities had no answer to the warfare Hezbollah was carrying out against the Israeli army stationed in Lebanon. The sort of guerrilla war that was carried out against Israel placed Hezbollah in an advantageous position over Israel, since they fought in an area that they know much better than the Israeli soldiers. Moreover, they were supported by the residents who gave them shelter, food and information.

Consequently, as the number of Israeli casualties grew, internal pressure on the government "to do something" to end the bloodshed, grew as well. Israel could have chosen to escalate the war and hit the infrastructure of Lebanon. However, it rightly feared that this would lead Hezbollah to launch massive amount of rockets and missiles against Israeli cities. All this, Israel feared, would bring Israel into an all-out war in Lebanon at a time when Israel was intensively occupied with the Intifada. Israel chose to withdraw from Lebanon believing that this would bring an end to the motivation of Hezbollah to continue the fighting: the reason was that an Israeli withdrawal from the southern part of Lebanon, and a consequent breaking up of the Israeli alliance with the Christian community in Lebanon, would end the motive of Hezbollah to continue its warfare against Israel: "I came to the conclusion," stated Ehud Barak in 2007," that our stay in Lebanon leads us nowhere, and we have to get ourselves out of Lebanon. I believed that our withdrawal from Lebanon would bring an end to the motivation of Hezbollah to fight against us. Moreover, Hezbollah is aware of the fact that our withdrawal from

Lebanon would give us full legitimacy to retaliate forcefully against Lebanon in general, and against Hezbollah in particular. All this will deter the Hezbollah from carrying out provocative acts against Israel."[24]

In fact the opposite happened: on October 7, 2000, Hezbollah kidnapped three Israeli soldiers who were on duty near the Lebanese border. Later on, it became evident that they died in this act of terror against Israel. At the same time, Hezbollah escalated its warfare against Israel, leading to major military confrontations in which Israel's main cities were attacked. Israel did not retaliate in a way that would have deterred Hezbollah. Eventually, the tension between Israel and Hezbollah led to the outbreak of an all-out war – The Second Lebanon War.[25]

The same thing happened in Gaza following the disengagement from Gaza in 2005. Prime Minister Sharon sincerely believed that Israel's withdrawal from Gaza and the destruction of all Israeli settlements there would end the motivation of the Palestinians in Gaza to continue their warfare against Israel. After all, once Israel withdrew from Gaza, the Palestinians would not be able to claim that Israel is occupying Gaza. In practice, the opposite happened, as demonstrated in the figure opposite, with a sharp increase in rocket attacks. The tension between the parties escalated, leading here too, to major military confrontations between Israel and Hamas.[26]

The present Israeli leadership firmly believes that there is no single reason to doubt that similar developments would occur following an Israeli withdrawal from the West Bank. Furthermore, an Israeli withdrawal from the West Bank would also have internal political repercussions. It would lead to the downfall of the present Israeli government. An Israeli withdrawal would involve the evacuation of thousands of Israelis from their homes in the West Bank. This would be accompanied by massive demonstrations in the state, which could escalate to violence and broad civil disobedience. The present coalition is clearly dominated by right-wing elements and would oppose any Israeli withdrawal from areas in the West Bank unless in the framework of a comprehensive peace agreement, which seems highly unlikely in the foreseeable future.[28]

As already mentioned, although Hamas is adopting extremely hostile positions towards Israel, its existence serves certain interests of Israel, from the point of view of the present Israeli government. The rule of Hamas in Gaza weakens the international pressure on Israel to implement the formula of the two-state solution. According to this formula, a peace treaty between Israel and the Palestinians would lead up to the establishment of an independent Palestinian state in the territories occupied by Israel during the Six Day War, namely the West Bank and the Gaza Strip.

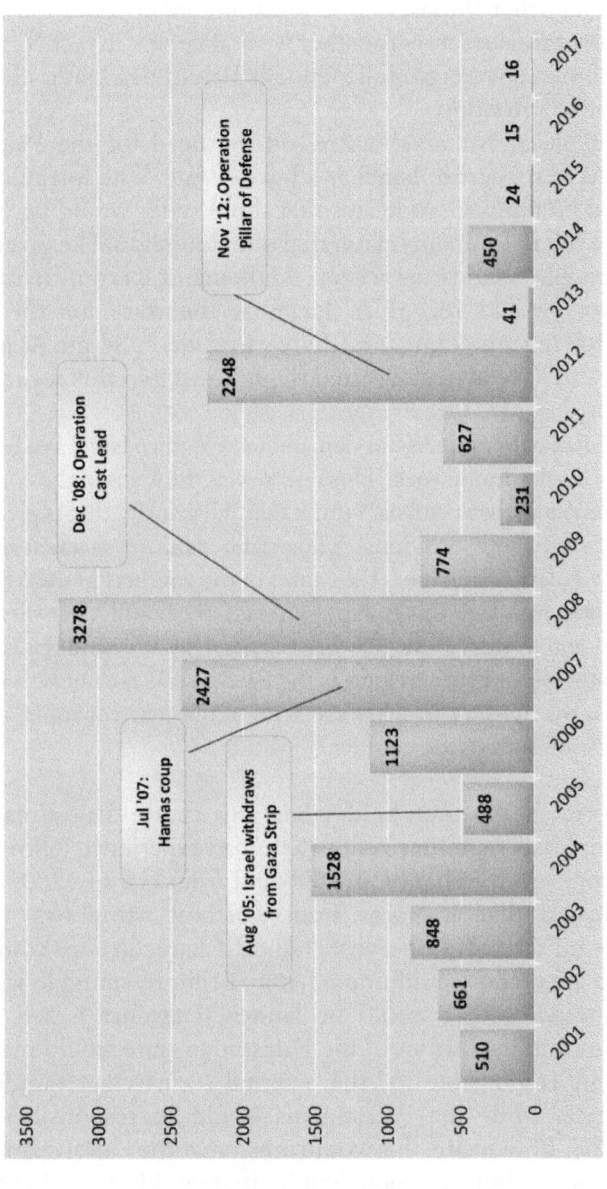

Number of rocket attacks annually from the Gaza Strip 2001 between 2001 and 2017.[27]

Prime Minister Netanyahu has accepted this formula in his famous Bar-Ilan speech in June 2009. However, in this speech he made clear that his agreement to the buildup of an independent Palestinian state would be based, among other things, on the following main conditions. When we combine all of those factors together we realize that in fact Netanyahu does not think of a sovereign, independent Palestinian state. He in fact refers to a sort of autonomy.

In the first place, Netanyahu stressed the need for the Palestinian Authority (PA) to recognize Israel as a Jewish state. The formula of two states, claimed Netanyahu, indicates that those states would be "for two people". Since the Palestinians claim that one state should be given to the "Palestinian people", and Israel accepts this demand, it is only natural that the Palestinians would recognize Israel as the state for the Jewish people.[29] Following a meeting with Italy's President, Sergio Mattarella, in November 2016, Netanyahu stated the following in this regard:

> The Palestinian entity rejects our demand to recognize Israel as a Jewish state, in any border framework. This rejection is the focus of the conflict between the Palestinians and us. I think that the priority that is given to the issue of settlements is wrong. Those who claim so, claimed in the past that the Palestinian issue is the center of the conflicts in the Middle East. We know now that this is wrong. In fact, the conflict between the Palestinians and us existed long before the Six-Day War and the rise of the settlements as an issue. As long as the Palestinians are not be willing to recognize the State of Israel as the state of the Jewish people, there will not be a solution to the conflict.[30]

Furthermore, Netanyahu holds that any future Palestinian state should be completely demilitarized. Israel's bad experience following the withdrawal from south Lebanon in 2000 and from Gaza in 2005 led her to the conclusion that any area from which it withdraws must be completely demilitarized. Otherwise, radical Islamic groups would take control of the area, and would immediately begin to smuggle weapons, rockets, and missiles that would be launched against Israeli civilian targets.[31] The demilitarization of the Palestinian state would mean that Israel would be responsible for the external security surrounding the Palestinian state, while the Palestinians would be responsible for its internal security. Practically, this would mean that the Palestinians would have no army units, but only police units that could carry "light arms" rather than "heavy arms." Furthermore, the Palestinian state would not be allowed to sign alliances with other states or to allow foreign armies to enter into their territory. Israeli and international forces would be responsible for monitoring the demilitarized state.[32]

Prime Minister Netanyahu explained that Israel cannot trust foreign soldiers to ensure that the areas it would evacuate remain demilitarized. After all, the security of the State of Israel is not their first priority, and this is understandable, taking into consideration that their loyalty is naturally given to their home countries. Israel has already seen how foreign soldiers flee their post once threatened by Hezbollah or Hamas.[33]

Moreover, Netanyahu's position demands that the Palestinian state will have to include the West Bank and Gaza. As long as Gaza is controlled by Hamas, the establishment of an independent Palestinian state in the West Bank would mean that there would actually be two Palestinian states. Israel cannot tolerate this situation, as it would pose grave risks to its very existence. The international community led by the United States will most likely accept this Israeli argument.[34]

Any agreement on a two-state solution, Netanyahu states, will have to ensure that the united city of Jerusalem will remain under Israel's sovereignty. Israel will not agree to a division of the city as proposed by the Palestinians. Israel will of course ensure the religious rights of all religions – Jews, Christians, and Muslims. However, this will not include de jure control over parts of the city.[35] The formal position of the Palestinians is that East Jerusalem is the capital city of the Palestinian state: "Jerusalem", stated the head of the Palestinian Authority only recently in the United Nations General Assembly, "is not for sale." He then harshly denounced President Trump's decision to move the American embassy to Jerusalem.[36]

Any agreement with the Palestinians, Prime Minister Netanyahu stated, would have to ensure that the conflict between the parties would come to its end. Israel will not tolerate Palestinian claims that an agreement between the parties will end only the "1967 conflict", while the "1948 conflict" remains unresolved.[37] If the Palestinians would not agree to a complete termination of the conflict, this would mean that even in the event of an Israeli withdraw from the occupied territories, the armed struggle against Israel would continue. In fact, only few months after the Oslo Agreement was signed, head of the Palestinian entity at the time, Yasser Arafat, disclosed to a group of Palestinians in Johannesburg, South Africa, that this was indeed his intention:

"The Jihad [Islamic holy war]", he stated, "will continue . . . Now after this agreement . . . our main battle (will come). Our main battle is Jerusalem. Jerusalem. The first shrine of the Moslems. This agreement, I am not considering it more than the agreement which had been signed between our prophet Mohammed and Koraish, and you remember the Caliph Omar had refused this agreement and [considered] it a despicable truce. We are in need of you as Moslems, as warriors of Jihad [in Arabic, Mujahideen]."[38]

Although he did not say it explicitly in his Bar-Ilan speech, Netanyahu often creates a connection between a settlement with the Palestinians and the removal of the Iranian threat. Any settlement with the Palestinians would necessarily entail grave risks to Israel's security. Israel is ready to undertake those risks for the sake of bringing an end to the long enduring Arab-Israeli conflict. However, Israel cannot afford itself to undertake such risks as long as the Iranian threat is still present.

The escalating tensions between Israel and Hamas on the one hand, following Israel's withdrawal from Gaza in 2005, and Hezbollah on the other hand, following Israel's withdrawal from southern Lebanon in 2000, has certainly been a traumatic experience for Israel, with wide implications on the formulation of its policies towards the Palestinians. Facing threats from two major terror organizations on its borders, Israel adopted a policy of containment, which was based on a defensive deployment of the IDF that would minimize the number of casualties in Israel, and would not compel Israel to undertake retaliatory operations that would necessarily escalate existing tensions in the region. Additionally, in case that Israel would have to retaliate, it would do so in a manner that would reflect her willingness to bring about an end to the escalation as soon as possible. In any case, Israeli retaliation was to be limited in its scope, and proportional to the provocation.

Israel pursued a ceasefire and de-escalation policy, estimating that Hamas and Hezbollah would be satisfied with limited confrontation and low level fighting, which would enable them to claim that they are the only body which is not afraid to challenge "the most powerful army in the Middle East"; namely – Israel. However, these measures proved fruitless, as Hezbollah and Hamas had other aspirations. They wished to prove again the validity of the statement made by Hassan Nasrallah, the head of the radical Islamic terror organization, Hezbollah, few days after Israel's hurried withdrawal from Lebanon in May 2000:

> The first victory we've made was liberating a big part of our land and a big number of the detainees in the occupation prisons. Besides, our jihad, resistance, steadfastness, and sacrifices defeated the enemy. We're here today enjoying freedom and safety, for the enemy's aircrafts do not dare fly in the airspace. I tell you so because "Israel" that really feared a wooden model of a Katyusha rocket launcher placed in Kfarkila is too coward to attack you on such a day! . . . An honest and serious resistance can make the dawn of freedom rise. Our fellow, beloved Palestinians, I tell you: "Israel," which owns nuclear weapons and the strongest air fleets in the region, is feebler than a spider web- I swear to God.[39]

Nasrallah made this statement regarding Israel's weakness as a result of Israel's containment policy vis-a-vis the violent measures undertaken against Israel by Hezbollah and Palestinian terror organizations. This policy was later on justified as a result of Israel's unwillingness to be engaged in warfare on two fronts – one in Lebanon, the other in Gaza and the West Bank (the Intifada).

However, beyond that calculation there were most likely other considerations that contributed to the adoption of the containment policy by Israel. Many people in Israel, within the security leadership and outside it, had grave doubts as to Israel's ability to achieve victory in such an asymmetrical confrontation. They tended to claim that while democratic societies can cope with other regular armies, they find it almost impossible to cope with terrorist organizations that operate mainly against civilian targets within Israel. In particular, they claimed, that Israel cannot cope with organizations that launch suicide attacks inside Israeli cities.[40]

The terrorists who carry out those actions look like ordinary citizens, and there is almost no legal way to prevent their entrance into crowded places. Moreover, there is a question on how it is possible to deter someone who wishes to die while carrying out his mission. What kind of threat can one make to someone ready to sacrifice his life for the cause that he believes in? Their conclusion was simple and clear: basically, we have to understand that there is no military solution to Palestinian terror against Israel. The only way to resolve this threat is through political agreement with the Palestinians.[41]

Hamas and Hezbollah naturally estimated that these statements and similar others made by top Israeli officials clearly singled Israel's weakness and fear of confrontation with terror organizations. This only encouraged them to escalate their operations against Israel. At a certain point, Israel could no longer tolerate the continuation of these provocations and it retaliated in a way that eventually led to a major confrontation.

There is no doubt that Hamas was aware of the likelihood of a major confrontation with Israel. There is also no doubt that Hamas knows very well that Israel's military abilities are overwhelmingly superior to those of Hamas. However, it rightly estimated that the ability of Israel to utilize all of its military capabilities is very low. Various external and internal factors, which will be examined in detail later on, would almost certainly prevent Israel from using its overwhelming military power in order to defeat Hamas.[42] While there was no doubt that Hamas would suffer losses, following this confrontation, Hamas would enhance its status within the Arab world, and in particular, within the different Palestinian organizations. Hamas estimated that the potential reward exceeded the costs of the risks that they were about to undertake.

That is the background of the events that eventually led to the outbreak of a major confrontation between Israel and Hamas in the summer of 2014. On July 8, 2014, Israel launched Operation Protective Edge to end terror against Israeli civilians in southern Israel, neutralize the threat from Hamas' vast tunnel network, and to reinstate deterrence.

During this operation, the Obama Administration affirmed its support for Israel's right to defend itself. However, it also adopted positions that in fact would not enable Israel to translate its military superiority over Hamas in order to inflict upon them a devastating defeat, which in turn would create a viable and strong Israeli deterrence.

The Obama Administration repeatedly and openly condemned Hamas' use of terror tactics. However, the administration also loudly pressured Israel to exercise restraint and proportionality, thereby introducing a political dynamic to the military operation and limiting Israel's military freedom of action.

CHAPTER

1

Hamas and Israel

Ever since Hamas seized control of the Gaza Strip in 2007, the terror organization has ratcheted up its pace of armament with rocket launching technology and sophisticated weaponry. This has destabilized the region and increased tensions with Israel. Since 2007, Israel and Hamas have engaged in a series of confrontations, including Operation Cast Lead (December 2008–January 2009), Operation Pillar of Defense (November 2012), and Operation Protective Edge (July–August 2014). Israeli citizens had no doubt that the overwhelming amount of weapons Hamas had been stockpiling was intended for use only against Israel. In 2014, Israel estimated that this stockpile numbered around 10,000 rockets, including Qassam, Katyushas, and GRADs.[1]

This chapter provides a brief background of Hamas' ideological and political aims, and examines the two earlier conflicts: Operation Cast Lead and Operation Pillar of Defense.

Hamas is the Palestinian branch of the Muslim Brotherhood that operates as an Islamic, militant, and sociopolitical organization. It regards Islam as an all-encompassing ideology, which can present a solution to every personal or national problem. It characterizes itself "by its deep understanding, accurate comprehension, and its complete embrace of all Islamic concepts of all aspects of life. This includes culture, creed, politics, economic, education, society, justice and judgment, the spreading of Islam, education, art, information, science of the occult and conversion to Islam."[2] Islam, Hamas stresses, informs all of Hamas' principles, policies, and calls to action.[3]

Hamas embeds nationalism within religion. It considers Palestine to be holy Muslim land, or *waqf*, and therefore "strives to raise the banner of Allah over every inch of Palestine."[4] Hamas views the creation of Israel as a Zionist-imperialist usurpation of the land and calls for its destruction. It perceives all peaceful solutions, initiatives, or conferences with Israel as tantamount to betrayal of the Palestinian people and their aspirations. According to its charter, "there is no solution for the

Palestinian question except through Jihad."[5] Hamas' charter is saturated in anti-Semitic rhetoric. It demonizes Jews, claiming that Zionists aim at "undermining societies, destroying values, corrupting consciences, deteriorating character, and annihilating Islam."[6] In fact, the Hamas charter quotes the following Hadith passage as a rallying cry: "The Day of Judgment will not come until Moslems fight the Jews, when the Jew will hide behind stones and trees. The stones and trees will say O Moslems, O Abdulla, there is a Jew behind me, come and kill him." Israel, the United States, the European Union, and Canada consider Hamas a terrorist organization due to "(1) its violent resistance to what it deems Israeli occupation of historic Palestine (constituting present-day Israel, West Bank, and Gaza Strip), and (2) its rejection of the off-and-on peace process involving Israel and the Palestine Liberation Organization (PLO) since the early 1990s."[7]

Hamas poses an immediate, ongoing threat to Israel, with those living nearby the Gaza border being particularly at risk. However, this is not an existential threat. At no point has Hamas ever put the survival of the State of Israel into question. Since its inception in 1987, Hamas has adamantly pursued the expansion of its military capabilities in targeting Israeli civilians. Funded by state-sponsors such as Iran and Syria, Hamas' ultimate goal has always been to destroy Israel and "liberate" Palestine. At a certain stage, Hamas succeeded in posing a threat to one of Israel's vital interests – the security of its air traffic. This occurred during Operation Protective Edge, when many air companies around the world canceled their flights to Israel for some days. This issue will be discussed in the third chapter.

Since Israel withdrew from the Gaza Strip in 2005, Hamas has exercised de facto rule over the Palestinian enclave. The relationship between Israel and Hamas can be defined as one of "fragile deterrence." Israel's military superiority over Hamas is significant, as Israel possesses the ability to target and destroy Hamas' vital infrastructure, weapons, personnel, and even its leadership. However, practically speaking, Israel has no real option to exhaust its military capabilities against Hamas due to internal and international constraints. In all of the confrontations between Israel and Hamas, Israel has made use of only a small percentage of it military capabilities.

Although there are long periods of détente between Israel and Hamas, Hamas typically uses these periods to expand and modernize its weapons arsenal before striking Israel again. In return, Israel meets Hamas' rocket attacks with retaliatory air strikes that seek to neutralize the threat and degrade Hamas' military capabilities. This cycle of détente and attack is seen in Operation Cast Lead, Operation Pillar of Defense, and Operation Protective Edge.

Operation Cast Lead

On December 18, 2008, Hamas stated that the ceasefire agreement between Israel and Hamas, which had been agreed upon on June 19, 2008, was no longer valid. It accused the "Zionist enemy" of failing to keep its commitments. Prime Minister Olmert stated that Israel had no interest in a war. Nabil Abu Rudeina, Mahmoud Abbas' spokesman, reflected upon the Palestinian Authority's opposition to Hamas' decision to cancel the ceasefire agreement with Israel. He stated that the ceasefire was essential for the Palestinian people. Additionally, Egypt's Foreign Ministry spokesman expressed deep concern over this decision, and implicitly blamed Hamas for the escalation. Egypt, he said, would be willing to negotiate with the conflicting parties for a renewed ceasefire if the parties would show readiness for such a dialogue. The United States Secretary of State, Condoleezza Rice, expressed America's hope that there would be no rise tensions near Gaza, and the State Department spokesman clearly blamed Hamas for the escalation.[8]

The launching of rockets from Gaza towards Israel intensified. It was clear that Hamas wished to create new rules of "the game." Notably, Hamas wanted Israel to ease its siege over Gaza, and to refrain from closing the supply to Gaza whenever rockets are launched towards Israel. This is despite the fact that Hamas had repeatedly launched rockets against the supply centers at the Erez and Kerem Shalom crossings. Israel was aware that if it permitted Hamas to achieve this goal, Israel's deterrence image would be severely jeopardized, and that Hamas' demands would only grow as they continued to launch rockets towards Israel. Prime Minister Olmert found himself to be in a complicated political position. His rivals claimed that he lacked the military and strategic experience to carry out a war against Hamas. In short, he had to prove his abilities in security affairs to the public. While visiting Qatar, Mahmoud Abbas (Abu Mazen) expressed his concern over the deteriorating situation in Gaza. He criticized Israel for its aggressive policy, but at the same time called upon Hamas to stop launching rockets towards Israel.[9]

The fact that the ceasefire between Israel and Hamas unraveled so quickly was not surprising. From its very beginning, many viewed the ceasefire as a half-hearted agreement intended to give each side time to build-up its capabilities in anticipation of a future conflict. Indeed, during the ceasefire, Hamas escalated its military build-up through weapons smuggling, while Israel reportedly devised plans for a Gaza offensive and made efforts to improve its defensive capabilities.[10]

The conflict began after Israel launched a raid to neutralize tunnels during the ceasefire on December 19. Shortly thereafter, Hamas issued a statement claiming that "the ceasefire is over, and there will not be a renewal because the Zionist enemy has not respected its conditions."[11] Hamas retaliated by launching more than 120 rockets into southern Israel between December 24 and December 25. The dramatic increase in Hamas' rocket fire was the apparent catalyst for Israel to launch Operation Cast Lead on December 27, 2008.

The Gaza conflict of 2008 served Hamas' political agenda. Following erosion in public support and trust, Hamas sought to rally popular support by "reclaiming the mantle of 'heroic resister."[12]

On December 18, 2008, Prime Minister Ehud Olmert delivered a speech at an annual conference held by Israel's Institute for National Security Studies (INSS). His main emphasis was on the Iranian nuclear threat and the need to make every effort to conclude a peace with Syria. He also referred to the threats posed against Israel by Hezbollah and Hamas, which are both supported by Iran. Finally, he argued against those who had claimed that the disengagement of Israel from the Gaza Strip was the reason for the rise of Hamas to power there, and for the deterioration of the security situation:

> There are those who claim that Hamas' takeover of the Gaza Strip resulted from the disengagement and made the security situation in the Gaza Envelope communities worse. I am convinced that were it not for the disengagement, the security situation in the South would be much worse; a large number of Israeli citizens would have been in the range of grenades and Molotov cocktails being hurled by residents of Gaza, and tens of thousands of soldiers would have to be in the heart of the Gaza Strip in order to protect its Jewish residents. Our memory at times can be short, but the Gush Katif communities before disengagement were attacked on a daily basis by mortars and rockets.[13]

On December 27, 2008, following the failure to extend an Egyptian-brokered six-month ceasefire that expired on December 19, 2008, Israel launched Operation Cast Lead in response to Hamas' increased rocket fire from the Gaza Strip. Israel's stated objective of the operation was to eliminate the rocket threat, weaken Hamas, and restore deterrence.[14] In 2005, Hamas launched 401 rockets toward Israel; in 2006 – 1,722; in 2007 – 1,276; in 2008 – 2,048; in 2009 – 566.[15] The sharp increase in the number of rockets launched against Israel in 2008 is of particular note. Furthermore, the rockets Hamas used were much more destructive, accurate, and long reaching than those used in the past. This had a severe negative effect on the morale of Israelis living in the Negev, as their security was now in even more jeopardy than before.[16]

Operation Cast Lead began with massive air strikes on various targets throughout Gaza. These targets included vital terror infrastructure,

Hamas bases, buildings that belonged to Hamas, and a Hamas TV station. Prime Minister Olmert made it clear that the operation would not end quickly. According to various reports, over 250 Palestinians were killed in these air strikes, and 750 Palestinians were reported wounded. Hamas leader Khaled Mashal stated, "Israel's aggression will not scare us. We will not let Israel impose on us its conditions for ceasefire. We demand the elimination of Israel's siege over Gaza." He also called upon Arab states to help Hamas during this difficult time, and appealed to the Palestinians in the West Bank to rebel against "Israel's occupation."[17]

On December 27, 2008, just before Operation Cast Lead began, Prime Minister Olmert delivered a speech setting out that over the past seven years, Hamas and other terrorist organizations had been indiscriminately attacking innocent Israelis in the southern part of the country, threatening the lives of thousands of civilians who wanted nothing more than to live peacefully in their homes, and to carry on their lives in a comfortable and normal way. Israel, he argued, did everything in its power to cooperate with the principles of the ceasefire and to de-escalate tensions. Unfortunately, these efforts were met by continuous attacks and violations of the basic understandings that were reached between Israel and Hamas with the assistance of Egypt. No country in the world, he said, can accept or acquiesce with this approach. It was clear that Hamas was not only prepared to continue its attacks, but to increase the launching of Qassam rockets and mortar shells against Israelis in the southern part of the country.[18]

According to a survey conducted by the Israel Democratic Institute, a solid majority of Israelis thought the war served Israel's best security interests.[19] The operation in Gaza, Prime Minister Olmert stated, "intends primarily to change the situation in the southern part of our country. It may take some time, and all of us are prepared to carry the burden and the pains that are an inseparable part of this situation. We did everything in order to make sure that Israelis in the southern part of the country will be protected under the circumstances. It's not going to be easy. It's not going to last just a few days. We are not fighting against the people of Gaza. I take this opportunity to appeal to the people of Gaza. As I have said several times in the past, you, the citizens of Gaza, are not our enemies. Hamas, Jihad, and the other terrorist organizations, are your enemies as they are our enemies. They brought disaster on you and they try to bring disaster to the people of Israel. It is our common goal to make every possible effort to stop them, so that we will be able to establish an entirely different type of relationship between them and us." [20]

"The efforts that we made today in our strikes in Gaza," Olmert noted, were focused entirely on military targets. We tried to avoid, and I think quite successfully, to hit any uninvolved people. We attacked only targets that are part of Hamas organizations. We will continue to make an effort to avoid any unnecessary inconveniences to the people of Gaza. We will make every possible effort," he promised, "to avoid any humanitarian crisis in Gaza. The people of Gaza do not deserve to suffer because of the killers and murderers of terrorist organizations. I am certain that the Israeli public is united behind the goals of this operation."[21]

Despite their limited successes, Hamas managed to retain fighters, weapons, and tunnels. According to the U.S. Congressional Service Report on Israel and Hamas: Conflict in Gaza (2008–2009), "Even if the IDF Gaza Coordination and Liaison Administration's claim that 750 Hamas fighters were killed is accepted, Hamas's estimated strength before the conflict was 9,000, meaning it may still have considerable manpower. Moreover, while Israeli airstrikes may have destroyed up to 80 percent of the estimated 300 tunnels on the border, some tunnels still are visibly in use and others are being repaired."[22]

Hamas leaders sharply deplored the Israeli operation in Gaza. They called it a "holocaust," and a "genocide." Some of them reflected their disappointment over the passive manner in which Arab states reacted to the war. Egyptian Foreign Minister, Ahmed Aboul Gheit, condemned Israel for its acts in Gaza. However, he also criticized Hamas severely for its actions that led to the conflict. He also noted that Hamas demonstrates cruelty to its own people, and refuses the transfer of wounded Palestinians to be treated in Egypt.[23] On December 31, 2008, Egypt's President, Hosni Mubarak, delivered a speech in which he made the following main points:

1 Israel should stop immediately its strikes in Gaza notwithstanding the question of who is responsible for the present conflict.
2 The different Palestinian factions should unite in these difficult times.
3 Condemnation of radical states, notably Iran and Syria, which take advantage of the Palestinian plight for their own selfish interests.
4 Egypt will continue to supply humanitarian aid to the people of Gaza.
5 Egypt will also try to renew the ceasefire in the area.[24]

Israel's initial air raid quickly expanded into ground operations, which were aimed at cutting off northern Gaza to prevent Hamas from supplying its fighters with weapons in southern Gaza. During the operation, Israel destroyed much of Hamas' infrastructure and support network while successfully targeting the upper echelon of its leadership. In this regard, it should be noted that Israel carried out a strike against the office of the Hamas leader, Ismail Haniyeh, and destroyed it.[25]

At the same time, Hamas also expanded its operations against Israel. On December 29, 2008, an Israeli woman was killed in Ashdod, while ten other people were reported wounded. In a visit to the Hatzerim Israeli Air Force base, Prime Minister Olmert met with pilots who carried out the air raids in Gaza: "The Israeli Air Force is our main strategic arm," Olmert stated. "Its standard of operation is probably the highest in the world. You give all of us the confidence that you can carry out every mission we will decide to undertake. And you will do it in the best way."[26] In early January 2009, Turkish Prime Minister Recep Tayyip Erdoğan visited the region, where he met with the leaders of Egypt, Saudi Arabia, Jordan, and the Palestinian Authority. Erdoğan presented to them his proposal to end the conflict, which included the renewal of the ceasefire between Israel and Hamas for one year, and the placement of regional forces in-between Israel and Gaza to prevent the launching of rockets against Israel.[27]

The US remained supportive of Operation Cast Lead throughout its duration. US President George W. Bush stated that he could understand Israel's willingness to defend itself against Hamas attacks. He clearly blamed Hamas for the outbreak of hostilities. Instead of taking care of the real needs of the people of Gaza, Hamas chose to launch rockets against Israel. The United States, he said, wishes to renew the ceasefire. But this should be a solid ceasefire so that hostilities would not be renewed soon after it was concluded. Any agreement should ensure that Hamas would not be able to launch rockets against Israel. President-elect, Barack Obama, also expressed his deep concern over the confrontation. However, he made it clear he would not interfere in President Bush's decisions before he entered office. The State Department spokesman stated that the United States was cooperating with other states to bring an end to the fighting and to establish a track that would ensure Israel's security. In the meantime, the spokesman stressed, Israel should ensure a minimum of civilian casualties in Gaza. The French President, Nicolas Sarkozy, came to the region and met with the leaders of Egypt, Israel, and the Palestinian Authority. Sarkozy blamed Hamas for the confrontation and called for an immediate humanitarian ceasefire.[28]

On January 8, 2009, US Members of Congress passed a unanimous consent in the Senate recognizing the "right of Israel to defend itself against attacks from Gaza and [reaffirming] the United States' strong support for Israel in its battle with Hamas, and [supporting] the Israeli–Palestinian peace process."[29]

This overwhelming support trickled down to the civilian sector as well. According to The Israel Project National Survey, 56 percent of Americans either strongly or "not so strongly" blamed the Palestinians for the conflict, while only 17 percent strongly or "not so strongly" blamed Israel; 41 percent of Americans blamed Hamas for the outbreak of violence, and 66 percent blamed Hamas for the ongoing humanitarian crisis.[30] According to a similar Pew Research Center poll, 40 percent of Americans approved of Israeli actions in Gaza, while only 33 percent disapproved.[31]

In order to enhance international support for Israel's operation in Gaza, Prime Minister Olmert met with the Secretary of NATO, Jaap De Hoop Shepherd. Prime Minister Olmert described the terrible situation in which Israelis living near Gaza had been facing over the past seven years: "Do you know any state in the world that would have accepted such a situation," he asked Shepherd. The secretary noted that he fully accepted Israel's right to defend its people. However, he also expressed his deep concern regarding the killing of innocent people. Prime Minster Olmert stated that Israel was doing everything it could to prevent a humanitarian crisis in Gaza, even during intensive fighting.[32]

Hamas, however, viewed the conflict as a major success in undermining Israel's global position. Israel's attacks on Gazan cultural, residential, and governmental sites, Hamas claimed, despite the fact that they were allegedly used for military purposes as part of Hamas' "victim doctrine," eroded its international reputation.

According to Gabi Siboni, Director of the Military and Strategic Affairs Program and Cyber Security Program at the Institute for National Security Studies and editor of the journal *Military and Strategic Affairs*, "Hamas regarded both the recommendations of the Goldstone Commission, which was established by the UN Human Rights Council to investigate Israel's actions during the operation, and the harsh international criticism of Israel's policies toward the Gaza Strip as a significant achievement. The continuing erosion in international public support of Israel's legitimacy to respond to rocket fire from the Gaza Strip has deepened the Hamas leadership's understanding of the potential of utilizing civilian casualties in the Gaza Strip as a powerful means in the balance of power between the resistance movement and Israel."[33]

Following several weeks of conflict, Israel successfully reinstated deterrence as it destroyed rocket-launching capabilities and weakened Hamas' military support network. According to the Congressional Research Service Report, Israel and Hamas: Conflict in Gaza (2008–2009), "Hamas politburo Chief, Khaled Mashal, admitted that he had been surprised by Israel's assault, its length, and its ferocity. That Hamas has not been responsible for rocket launches since the ceasefire, even though it retains the capability, and did not directly engage the IDF, suggests that it may be reluctant to test Israel's will. Moreover, Israel's military achievements during the conflict were made at minimal physical cost to itself: 13 dead. The Palestinian casualties, according to the United

Nations, included over 1,400 dead and roughly 5,400 wounded, plus huge infrastructure and physical losses. The massive devastation may serve as the ultimate deterrent." [34]

The head of Israel's military intelligence branch at the time, Amos Yadlin, stated on January 11, 2009, that there was a big disagreement within Hamas on how to proceed with the conflict. While the leaders of Hamas in Damascus adopted stubborn positions, the leaders of Hamas in Gaza supported a compromise. Hamas was greatly surprised by the Israeli retaliation, and was very much worried about interior population support for Hamas. In addition, Hamas suffered from a lack of arms. This explained the decrease in the launching of rockets against Israel. However, Amos Yadlin explained that Hamas would not surrender, and remained capable of continuing the fighting.[35] Shortly before the end of the conflict, Israel and the United States signed a memorandum of under-standing on cooperation against the smuggling of arms into Gaza.[36] Thus, on January 17, 2009, Israel announced a unilateral withdrawal, with Hamas declaring a ceasefire shortly after.[37]

As the operation was coming to its end, Prime Minister Olmert summarized its results. Hamas, he stated, was severely harmed in the war, and its infrastructure was badly damaged. "We destroyed many of the tunnels that Hamas built in order to smuggle arms into Gaza. Hamas leaders and officials are hiding under the ground, fearful of being killed by us. We have made use of our impressive capabilities in the air, sea, and on the ground. This campaign has sharply enhanced Israel's deterrence image. Throughout the war, the Israeli people had demonstrated their resilience, notwithstanding the hundreds of missiles and rockets launched on Israeli cities and populated areas. Throughout the operation, we enjoyed the support of the international community, and many states supported our position that the armament of Hamas should be stopped, or at least significantly limited. During the years that preceded the war, Israel had tried to show restraint and to keep the profile of the fighting at a low level. Hamas interpreted this as a sign of weakness. Hamas learned the hard way that this was a grave mistake on their part. Notwithstanding the fierce fighting, we have not stopped providing humanitarian aid to Gaza. We also tried our best to avoid the killing of innocent civilians. We now believe we have achieved most of our purposes in the war. It's time to bring an end to the fighting."[38]

One day later, on January 18, 2009, it became evident that Prime Minister Olmert's statement regarding the support Israel enjoyed from the international community was accurate. In an unprecedented gesture, the leaders of Europe came to Israel in order to demonstrate their sym-pathy and support. Among them were: President of the European Council, Prime Minister of the Czech Republic, Mr. Mirek Topolanek;

President of France, my friend Mr. Nicholas Sarkozy; Chancellor of Germany, Ms. Angela Merkel; Prime Minister of Great Britain, Mr. Gordon Brown; Prime Minister of Italy, Mr. Silvio Berlusconi; Prime Minister of Spain, Jose Luis Rodriguez Zapatero.[39] In his opening statement, Prime Minister Olmert stated that the Israeli people wish to express their appreciation to the leaders of the European countries for demonstrating their impressive support for the State of Israel, and concern for its safety. He told them, "The united front which you represent and your uncompromising stand with regard to the security of the State of Israel warms our hearts and strengthens us at this sensitive time."[40]

Prime Minister Olmert told his visitors that Israel decided to launch a military operation in order to thoroughly change the reality in the southern communities with regard to security. The reality in which the residents of Sderot, the Gaza Envelope and other communities had lived for many years was an impossible one – intolerable for these citizens, and impossible for the nation. He concluded, "No sovereign nation would allow its civilians to be harmed; no enlightened regime suffers indiscriminate fire directed at its residents."[41]

"Today," Prime Minister Olmert added, "after three weeks in which the IDF and security services conducted an outstanding military operation and struck a serious blow to the Hamas organization, and after we realized the goals we determined as we launched the campaign – we decided on a ceasefire. We did so to comply with the request of the Egyptian President, Hosni Mubarak. If the ceasefire is stable, especially in light of the statements we heard today," he promised, "the State of Israel had no intention of staying in the Gaza Strip. We are interested in withdrawing from the Gaza Strip as quickly as possible the moment we are assured that the ceasefire is being respected and is stable, and that there is no threat to the security of southern Israel."[42]

"Today," Olmert said, "it is clear to everyone that in order to achieve a stable ceasefire, Hamas must be prevented from building up its military capabilities through massive weapons smuggling from Iran and Syria to the Gaza Strip." "In the letter I received from you yesterday," Prime Minister Olmert told his visitors, "you expressed profound commitment to assisting in every way possible in order to ensure that weapons will not succeed in reaching the murderous organizations in Gaza. Now, we must translate that commitment together with Egypt and the United States, with whom we signed a memorandum of understanding on this matter, into actions, which will prevent the terrorist organization, Hamas, from rearmament. This is in the supreme interest of all those who fight the forces of evil. It is also in the interest of all those who believe wholeheartedly in and wish to advance the peace process between the Palestinians and us."[43]

"My Government," Prime Minister Olmert argued, "placed the matter of negotiations with the Palestinians at the top of our agenda alongside our concern for the security of Israel. We hope that stability in Gaza, the ceasefire, and the undermining of the Hamas regime that is the inevitable result of the strengthening of President Abu Mazen, will allow us to advance the peace process between Israel and the Palestinians as rapidly as possible. We have done, and will continue to do all that is necessary to prevent a humanitarian crisis in the Gaza Strip, and so that we can help the innocent civilians who fell victim to the terrorist organization."[44]

When Operation Cast Lead ended, the assessment was that Israel had made several achievements. First, it was assumed that Israel succeeded in its endeavor to convince Hamas that the rules of "the game" that had prevailed until then were no longer valid. Israel had carried out a disproportional response towards Hamas, thus bringing them to the belief that even if they wished to keep a low level profile during the fighting, Israel would not necessarily respond accordingly. Furthermore, Israel had most likely convinced Hamas that populated areas and civilian institutions, even mosques, would not enjoy any immunity from Israel's strikes. Hamas fighters will not be allowed to find shelter behind civilian occupied places.[45]

Furthermore, by the end of the operation, it was agreed to establish a system for inspection in order to prevent the smuggling of arms into Gaza. Though nobody believed that this system would prevent absolutely the transfer of arms into Gaza, it was believed that it would most likely decrease the scope of the smuggling. Finally, the war had revealed the sharp divide between moderate and extremist forces in the Arab world. For many years, the strength of the Arab world within the international community derived among others, from its perceived unity and solidarity. However, it would become apparent that the Arab world is far from being a unified force. This weakened the position of the Arab world within the international community while simultaneously strengthening Israel's position therein.[46]

Operation Pillar of Defense

On November 14, 2012, following an upsurge in cross-border violence between Hamas and Israel, Israeli Prime Minister Benjamin Netanyahu announced the launch of Operation Pillar of Defense. Many people in Israel   concluded that Hamas was no longer deterred from attacking Israel. This assessment was based on the following main components:

1   The rise of the Muslim Brotherhood in Egypt and the strengthening of relations between Hamas and Egypt.
2   Hamas' growing weapons arsenal that included long-range rockets with the potential to strike the center of Israel.
3   Hamas' decision to "loosen the reigns" and allow other armed groups to operate within Gaza.
4   Hamas' increased cross-border rocket fire targeting Israeli civilians.

Israel's stated goal was to neutralize the threat of rocket fire by targeting Hamas military capabilities, which would protect Israeli civilians and restore deterrence. There were also speculations of a hidden Israeli agenda to undermine, or possibly even destroy, the Hamas government. Toppling the Hamas government, it was claimed, would damage not only current military capabilities, but also future capabilities and aspirations.[47] However, it is doubtful that this assessment represented the interests of the Israeli government at the time. That is, Israel's goal was to inflict heavy damage upon Hamas, but not to eliminate it.

According to a survey conducted by the Israel Democracy Institute, a solid majority of the Israeli public, over 80 percent, supported the war.[48] The Operation began with a surgical air strike targeting Ahmed Jabari, head of Hamas' military wing in the Gaza Strip, accused of orchestrating "all terrorist activities against Israel from Gaza, including the kidnapping of Gilad Shalit." At the same time, the strike was also directed against the long-range missiles of Hamas, the Iranian Fager 5. [49] Hamas announced that this action "opened the gates of hell,"[50] and in retaliation dramatically increased its rocket attacks on Israel, including the central cities of Rishon LeZion, Holon, Tel Aviv, and Jerusalem. The IDF reported that between November 14 and November 21, 2012, Palestinian terrorist groups launched over 1,500 rockets against Israel.

On November 20, 2012, Prime Minister Netanyahu met with the Secretary of the United Nations. In a press conference following the meeting, Netanyahu made the following main points:

1   Israel appreciates the Secretary's insistence on Israel's right to defend itself and his condemnation of the launching of rockets towards Israel.
2   Israel joins the Secretary's concern over the killing and wounding of innocent civilians from both sides.
3   Israel should be praised for the efforts it undertakes to prevent harm to civilian population. No other army in the world takes such measures as the IDF does.
4   Hamas is targeting its strikes directly against our civilians. At the

same time, they hide their arms in populated areas. Thus, they are carrying out two war crimes.

5   There is no moral comparison between Hamas and the victims of its terror.

6   Israel will join the efforts to stop the fighting by political means.[51]

Later on that day, PM Netanyahu met with the Foreign Minister of Germany, Guido Westerwelle. Netanyahu praised Germany for its solid support of Israel in its battle against Hamas. Israel, Netanyahu alerted, wants to end the war by political means. It believes that Germany can play a key role in this regard. However, if a political solution is not possible, Israel will undertake all means to defend its people: "We will not tolerate the firing of rockets on our cities and people." The German Foreign Minister made clear that Germany stands by Israel. "We have to try to bring about a ceasefire. However, the first condition is that Hamas would stop launching rockets against Israel. This position is supported by all states in the European Union. We have also to bring Egypt into the political efforts for solution."[52]

At another event, Netanyahu set out the difference between Israel and Hamas. "We are doing our utmost to refrain from harming the civilian population. When, however, accidents happen and Palestinian civilians are killed, we see this as a failure on our part. Hamas, on the other hand, sees every killing of an Israeli as a success, and they make it clear that they are happy with this outcome. We live in a highly unstable region," Netanyahu said. "We have to be very careful in our decision-making. Behind Hamas terror stands Iran, which wants to annihilate the State of Israel. For this, it wants to develop a nuclear capability. We will not permit this to happen."[53]

The Israeli operation successfully targeted more than 1,500 identified terror sites in Gaza, including 30 senior Hamas and Islamic Jihad terrorists. In order to avoid as much as possible the killing of innocent civilians, Israel dropped pamphlets from the air to the residents of certain neighborhoods in Gaza, calling on civilians to leave their homes so that they would not be hit by the impending Israeli strike. On various occasions, Hamas called upon residents to burn those "Zionist papers." In many cases, Hamas tried to prevent residents from leaving their neighborhoods.[54] The operation lasted eight days, from November 14 to November 21, 2012, until Egypt brokered a ceasefire.[55]

The Egyptian-brokered ceasefire agreement between Israel and Hamas was highly controversial. Some claimed that the diplomacy gave Hamas international legitimacy and sent a mixed-message regarding deterrence. According to Avner Golov, Research Fellow at the Institute for National Security Studies, "This operation alone cannot completely restore the element of deterrence. While Hamas may have suffered a severe blow, the armed conflict enabled Hamas to extract concessions from Israel that it had been unwilling to consider in the past, such as an agreement to ease restrictions at the border crossings. Alongside the message of a heavy price, therefore, Israel has conveyed the message that escalation provides Hamas with significant leverage and benefits."[56]

However, others argue that diplomatic agreements actually benefit Israel's security interests. It convinces Hamas that Israel is not interested in a military conflict. However, if it were compelled to act, it would ensure that a military confrontation would cost Hamas more than they would be ready to tolerate. Giora Eiland, former head of the Israeli National Security Council, stated, "The more the Hamas government is required to meet the standard of state-like responsibility, and the more the economic situation improves and the construction of civilian infrastructures increases, the more the government in Gaza will be restrained in attacking Israel."[57]

Following the ceasefire agreement, Prime Minister Netanyahu admitted that there was disagreement within the coalition with regards to the ceasefire. Hamas, Netanyahu stated, estimated that Israel did not wish to engage in a major military confrontation. "Hamas was wrong. We have dealt a severe blow to Hamas. Many of its officers were killed, and large parts of its infrastructure were destroyed. All of this was done with the support of the international community. I wish to reflect my special gratitude to President Obama for his sweeping support in Israel at this period of time, and for allocating important resources for the advancement of the iron dome system. In my talk today with President Obama, we agreed it was the appropriate time to give the ceasefire a chance. We agreed to cooperate in an intensive effort to prevent the smuggling of weapons to Gaza. We are facing complicated situations, and we must act with great caution and responsibility, taking into account a variety of political and military considerations. I know there are those who think we should have escalated the fighting against the Hamas. We believe that at this stage, we had to stop the fighting and agree to a ceasefire proposal. Maybe in the future we will have to act in a more vigorous way. Now, we have to agree to end the fighting."[58]

Following the conclusion of the agreement, the White House announced that President Obama wished to reflect his appreciation to the leaders of Israel and Egypt for their efforts to bring an end to the fighting. The United States, President Obama added, would enhance its efforts to support Israel's security needs. In particular, the United States would enhance its efforts to prevent the smuggling of arms into Gaza. The US would also try to increase the financial support for the Iron Dome program. President Obama also called on Egypt's President Morsi, and they agreed that further efforts should be made to find a comprehensive solution to the Gaza crisis.[59]

Khaled Mashal stated that Hamas would keep its commitment to the agreement as long as Israel kept its own. Hamas, he stressed, had won a great victory over Israel. Mashal further praised the Egyptian leadership for standing in support of Hamas, and applauded Iran for its generous provision of funding and arms. Finally, he called upon the Palestinian Authority to admit that "resistance" was the only way to solve the Palestinian problem. In order to advance this cause, all of the Palestinian factions should be united.[60]

Although there was a brief period of peace and stability following Operation Pillar of Defense, it did not last long, as Hamas' main ally in Egypt, the Muslim Brotherhood, was forcibly removed from power in a military coup. Following this development, Hamas found itself to be increasingly isolated in the Arab world. Furthermore, Hamas' domestic popularity was crumbling. According to Gabi Siboni, "The economic system it had developed through the tunnels in the Rafah region was almost completely paralyzed by the countermeasures implemented by the Egyptian military. The sense of isolation and the desire to change the problematic position in which it now found itself is, most probably, what led Hamas to the most recent round of fighting."[61] Deterrence and détente once again eroded, eventually giving way to conflict.

CHAPTER
2

Operation Protective Edge

This chapter analyzes both the ultimate and proximate causes of Operation Protective Edge. The ultimate cause, or root cause, was Hamas' attempt to reclaim its eroding power by uniting its people against Israel. The proximate cause, or immediate catalyst, was the kidnapping and murder of three Israeli teens and the subsequent escalation in violence on both sides.

By the summer of 2014, Hamas was politically isolated and economically destitute. Furthermore, leading up to Operation Protective Edge, Hamas found itself with few Arab allies. During the Arab Spring, Hamas supported Sunni rebels against Syria's Bashar al-Assad, Iran's long-time ally. In retaliation, Iran coordinated with Syria and Hezbollah to cut off aid to Hamas, which was "one of Hamas' most vital lifelines."[1] The extent of the reduced aid going to Gaza was evidenced by Hamas' return to the use of low-grade, homemade rockets instead of high-quality rockets smuggled in from Iran.[2]

ISIS' rise also contributed to Hamas' sense of political isolation within the Arab world. According to Gabi Siboni, "In this context, the United States and the countries of the West suddenly found themselves on the side not only of Saudi Arabia and Jordan, but also of Iran, Hezbollah, and even the regime of Bashar al-Assad in Syria. This phenomenon, which may not guarantee the restoration of Washington's relevance to the events in the Middle East, has pushed Hamas and the problem of the Gaza Strip onto the sidelines of the international agenda, thus exacerbating its isolation even further."[3]

Hamas was also economically isolated as a result of Israel's crippling seven-year blockade of the Gaza Strip. According to the World Bank, half of all Gazan residents live in poverty and one in three are unemployed.[4] This led to the creation of Gaza's "tunnel economy," as Gazans illegally smuggled goods in from Egypt. However, Gaza's dire economic situation was then exacerbated by the Egyptian government's crackdown on smuggling tunnels and the closure of the Rafah crossing, the only crossing point between Egypt and the Palestinian enclave. Nearly 95 percent of tunnels to Gaza from Egypt were destroyed as part of these efforts.[5] Egyptian President Morsi issued these measures after declaring the Muslim Brotherhood a terrorist organization and accusing Hamas of contributing to terrorism in the northern Sinai.[6] The loss of income from tunnel taxes left Hamas unable to pay 42,000 civil servants, while unemployment rose to 46 percent (58 percent for those of working age under 30); per capita GDP was at an average of around $4 per day.[7]

At the same time, the Egyptian government was also in the process of establishing a one-kilometer buffer zone with Gaza to curb smuggling, which benefitted terrorists in the region. This involved the demolition of homes along the 13-kilometer Gaza border, displacing up to 10,000 Egyptian citizens.[8] Middle East strategist David Butter argues that "certainly from the Hamas point of view, the desperation they were in may have driven them to risk getting involved in [the escalation of violence preceding Operation Protective Edge] on the assumption that a major crisis would result in a major reappraisal of the entire economic condition." [9]

Finally, the collapse of diplomacy further motivated Hamas to pursue military means. Under pressures caused by increased political isolation and popular demand, Hamas and Fatah established a national unity government in June 2014. Hamas allegedly consented to the national unity government's three conditions: recognition of Israel, continuation of previous agreements, and the end of violence.[10] However, Israel responded to this announcement by walking out of peace talks, forcing Fatah to choose between peace with Hamas and peace with Israel. It chose the latter and aided the IDF in cracking down on Hamas following the kidnapping of the three Israeli teenagers. The short-lived unity government between Fatah and Hamas disintegrated. For Hamas, diplomatic solutions, concessions, and national unity did not improve its situation.

According to Dr. Jeroen Gunning, Executive Director of the Durham Global Security Institute, "Against this backdrop, Israel and Fatah's clampdown on Hamas, so soon after the establishment of a national unity government, signaled the futility of a political route at this stage, thus strengthening those favoring a military response."[11]

Given the deterioration of deterrence, Hamas' political and economic isolation, and the collapse of diplomatic solutions, Hamas believed that it had nothing to lose in attacking Israel. For Hamas, the status quo was so unsustainable that any action would have been an improvement. Therefore, when violence escalated following the murders of the teenage boys, Hamas was ready to re-engage.

Abduction and Murder of Three Israeli Teenagers

On June 12, 2014, Eyal Yifrach (19), Gil-ad Shaar (16), and Naftali Fraenkel (16) were kidnapped on their way home from a religious school (yeshiva) at a Gush Etzion bus stop around 10:00 p.m. Israeli authorities presumed that the three teenagers were either lured or forced into a car and taken south to the vicinity of Hebron, a large Palestinian city and Hamas stronghold. At 10:25 p.m., one of the teenagers called the police emergency numbers and whispered, "We've been kidnapped." Police officials assumed that this was a prank and dismissed the legitimacy of the claim, delaying the search for several hours.

Shortly thereafter the teenagers were confirmed as missing. On June 15, 2014, Israel's security authorities launched Operation Brother's Keeper to "find the three boys, their kidnappers, and to weaken Hamas infrastructure in Palestinian cities and villages including Hebron, Nablus, Ramallah, and Jenin."[12] The IDF arrested more than 400 Palestinians, including many of Hamas' top leaders.[13] However, many Israelis already feared the worst. The recorded call to the police had clearly indicated that the kidnappers shot the three students. Furthermore, in the West Bank, Palestinian kidnappers typically kill their victims shortly after abduction out of fear that they will be detected. The lack of ransom, credible claims of responsibility, or any other signs that the teenagers remained alive hinted at this tragic outcome.

In a speech delivered on June 14, 2014, Prime Minister Netanyahu stated that Israel's security forces were doing their utmost to find the teenagers. "The boys," he said, "were kidnapped by a terror organization." Netanyahu wished to emphasize that Israel did not accept the view that this operation was carried out by individual persons who were not members of Hamas. "Israel will take all of the necessary measures to prevent the transfer of the boys to Gaza." "Israel's security forces," Netanyahu said, "have to be ready for every development." Here, Netanyahu clearly hinted that Israel might retaliate and that this might lead to an all-out war. "Israel demands that the Palestinian Authority do everything [in its power] in order to bring the boys back home," he said. "The kidnapping was carried out in the West Bank, and it is under the responsibility of the Palestinian Authority. This kidnapping confirms what we have said many times: namely that the cooperation between the Palestinian Authority and Hamas only enhances violence in the region. All terror organizations wish to eliminate Israel and they don't need any excuse for that. It is unacceptable for us that the Palestinian Authority wishes to establish a joint government with Hamas, and at the same time wishes to proceed with the peace process. We live in a turbulent region and we can rely only on ourselves."[14]

A day later, Netanyahu again stated he holds the Palestinian Authority responsible for the kidnapping of the boys. "The fact that this violent act was carried out in an area under Israeli control does not matter. Just as we hold the Palestinian Authority responsible for act of terror in Tel Aviv, we can hold the Palestinian Authority responsible for this act, which had been carried out in the West Bank." Thus, in the first stages of the crisis Netanyahu put the blame almost solely on the Palestinian Authority. Later on, this viewpoint would change.[15]

On June 15, 2014, American Secretary of State John Kerry strongly condemned the kidnapping of the three Israeli teenagers: "Our thoughts and prayers are with their families," he stated. "We hope for their quick and safe return home. We continue to offer our full support for Israel in its search for the missing teens."[16] At the same time he wished to ensure that the cooperation between Israeli and Palestinian security authorities would not be damaged by this event: "We have encouraged full cooperation between Israeli and Palestinian security services. We understand that this cooperation is ongoing."[17]

Secretary Kerry further wished to emphasize that Hamas rather than the Palestinian Authority should be held responsible for the act: "We are still seeking details on the parties responsible for this despicable terrorist act," he explained, "although many indications point to Hamas' involvement. As we gather this information, we reiterate our position that Hamas is a terrorist organization known for its attacks on innocent civilians and which has used kidnapping in the past."[18]

In a speech delivered on June 16, 2014, there were clear signs that Prime Minister Netanyahu decided to shift from his previous position and to put the blame primarily on Hamas in Gaza. "We are dealing," he said, "with a very severe event, which will have very severe repercussions. We are undertaking measures against Hamas. We have arrested more than 100 persons connected with Hamas. At the same time, Hamas in Gaza goes on launching rockets against Israel. We will retaliate forcefully to this provocation." Netanyahu also announced that he had spoken with Secretary of State John Kerry, and expressed his appreciation for Kerry's condemnation of Hamas and his support for Israel's right to self-defense. "Israel expects other states to join in condemning this brutal act of terror," Netanyahu stated.[19]

During a meeting the following day with Tony Blair, Netanyahu continued to place blame on Hamas. "This act of terror," Netanyahu said, "exposes the real character of Hamas. The international community must deplore this act of terror. It has to call upon the leader of the Palestinian Authority to bring an end to its alliance with Hamas. You cannot speak about peace and at the same time collaborate with a regime that kidnapped children."[20] During a meeting with mayors from all over the world who came to Israel, Netanyahu explained that Israel's security forces were engaged in combat against Hamas, a terrorist organization determined to eliminate the State of Israel.[21]

On June 19, 2014, Netanyahu directly accused Hamas of kidnapping the three teenagers. He announced, "We're doing everything in our power to bring back our three kidnapped teenagers. They were kidnapped by Hamas. We have no doubt of that. It's absolutely certain."[22] "We are carrying out acts against Hamas in the West Bank," he said. "We arrest their people, we close their institutes, and we confiscate their money."[23] Netanyahu reflected this view a day later in a meeting he held with the parents with the three teenagers.[24]

This statement clearly indicated that Israel held the Hamas authorities in Gaza responsible for the brutal murder of the teenagers, and that they would be punished for it. In a speech delivered on June 26, 2014, Netanyahu stated that the kidnapping of the three teenagers demonstrated to Israel who her enemies were. "These are terrorists who are filled with great hatred to us. They have no moral restraint in their warfare. They just want to wipe us off the map. We will find the murderers and bring them to justice. We will carry out the struggle against Hamas with intensive efforts to refrain from inflicting harm to innocent people. Hamas will have to know that if we do not live in peace, they also will not live in peace."[25]

Hamas neither confirmed nor denied involvement; however, Hamas' political leader Khaled Mashal made the following points in a lengthy interview with *Al-Jazeera*:

1   The kidnapping of three Israeli teenagers is a heroic act.
2   No one claimed responsibility so far. I can neither confirm [Hamas's responsibility] nor deny it.
3   The kidnapped teenagers were "settlers and soldiers in the Israeli army."
4   Blessed be the hands that captured them. This is a Palestinian duty, the responsibility of the Palestinian people. Our prisoners must be freed: not Hamas' prisoners – the prisoners of the Palestinian people.
5   They [the three youths] are combatants. Settlers in the West Bank are a disaster . . . they burn agricultural produce, kill children by running them over, invade homes, burn mosques and attack churches.
6   The one who lost these three [youths] is Netanyahu by ignoring the Palestinian suffering and provoking our people.[26]

In a speech delivered at the end of June 2014, Prime Minister Netanyahu tried to place the kidnapping of the teenagers in a broader perspective – related to the dramatic revolutionary changes that were taking place in the Middle East, and the strategic challenges Israel was facing as a result of these changes. "The secular dictatorial regimes that had been established after the end of the colonial regimes have fallen, and brought about an unprecedented level of violence in our region. The hope that liberal powers will take the lead in the region has evaporated and radical Islamic movements [have taken] the power. This turmoil will go on for many years. It poses in front of us several strategic challenges: in the first place we will have to make every effort to defend our borders."[27]

"The radical Islamic groups," Netanyahu stated, "are already operating near our borders in the north and in the south. In Lebanon and in Syria, Hezbollah and Iran are posing grave dangers to us. In the south, ISIS is gaining power in Sinai and Hamas is also enhancing its power with massive support from Iran. We will have to build a sophisticated fence along our borders. This fence will not prevent the launching of rockets against us, but it will drastically limit the scope of infiltration into Israel. We will also insist that our border on the eastern side will be on the Jordan River. Nobody can predict the future developments in Syria, Iraq, and Jordan, and we have to be ready for the worst to come. We will also have to bolster our control over the West Bank. We cannot rely on anyone but our own security forces to ensure our security. Our friends believe our security could be ensured by international forces or by local forces trained by Western states, in particular the United States. We do not believe in the validity of these assessments. The present turmoil in the Middle East opens to us new opportunities to enhance our relations with moderate Arab states especially in the Gulf. We are facing common enemies and common threats, and we all believe that no external power can be trusted to ensure our security. Finally, we have to make every effort to ensure that Iran does not become a nuclear power. If we fail and Iran gains a nuclear capability, our very existence will be threatened."[28]

Following an intense 18-day search, the bodies of the three teenagers were discovered on June 30, under a pile of rocks in an open field about 15 miles from the Gush Etzion bus stop where they had been last seen. Prime Minister Netanyahu and his wife Sarah visited the families of the murdered teenagers on July 6, 2014.[29] At the funeral of the three boys, Prime Minister Netanyahu set out the huge moral gap between our enemies and us. "We cherish life and they cherish death, we cherish mercy and they cherish cruelty. These are the secrets of our power. These are the values that enable us to defeat our enemies."[30]

The kidnappings engulfed the nation. According to the *New York Times*, "as news spread, Israeli television channels halted World Cup broadcasts and canceled prime-time shows, filling the hours with discussions of the discovery, while radio stations played sad songs. People converged in Rabin Square in Tel Aviv, sitting on the ground and lighting candles. Others recited psalms at the West Bank hitchhiking post where the teenagers were last seen."[31] While the nation mourned, the government spurred into action. In an emergency cabinet meeting, Prime Minister Netanyahu announced, "[the teenagers] were kidnapped and murdered in cold blood by beasts. Hamas is responsible, and Hamas will pay."[32] Shortly afterwards on June 30, 2014, President Obama issued a statement: "On behalf of the American people," he avowed, "I extend my deepest and heartfelt condolences to the families of Eyal Yifrach, Gilad Shaar, and Naftali Fraenkel – who held Israeli and American citizenship. As a father, I cannot imagine the indescribable pain that the parents of these teenage boys are experiencing. The United States condemns in the strongest possible terms this senseless act of terror against innocent youth. From the outset, I have offered our full support to Israel and the Palestinian Authority to find the perpetrators of this crime and bring them to justice . . . As the Israeli people deal with this tragedy; they have the full support and friendship of the United States."[33]

At the same time, President Obama sent a clear message to Israel that this incident should not lead Israel to carry out retaliatory acts which might be harmful to the cause of peace: "I encourage Israel and the Palestinian Authority to continue working together in that effort. I also urge all parties to refrain from steps that could further destabilize the situation."[34] By directing his appeal to both Israel and the Palestinian Authority, President Obama clearly indicated that he did not agree with Israel's claim that Palestinian incitement was a major cause for acts of terror carried out against Israelis. Nevertheless, despite urges for restraint and de-escalation, the situation deteriorated.

Abduction and Murder of Mohammad Abu Khdeir

On July 2, three Israelis (two unnamed minors and Yosef Chaim Ben David) kidnapped Mohammad Abu Khdeir, a 17-year-old Palestinian teen, and then beat him and burned him alive in an apparent revenge killing. Mohammad Abu Khdeir was abducted outside of a mosque near his home in East Jerusalem. Witnesses said he was abducted, dragged into a car by several men, and driven away. His body was found later that day, "burned beyond recognition,[35] in the Jerusalem Forest. Jerusalem police immediately arrested six suspects, three of whom confessed to the murder. Soon afterwards the editor of the Hamas journal stated that

Israel had in fact adopted the Nazi ideology and is carrying out a holocaust against the Palestinian people.[36]

Netanyahu condemned the murder as a "heinous" act, saying, "Such murderers have no place in Israeli society. I promise you that we will bring full force of law to bear against the perpetrators of this horrific crime, which deserves every sort of condemnation and rejection . . . we won't distinguish between [Palestinian] terror and [Jewish] terror, and will deal severely with both. I don't distinguish between incitement and incitement in the state of Israel."[37]

He added that, "this is a difference between us and our neighbors. There, murderers are received as heroes, and city squares are named in their honor. That isn't the only difference between us. The inciters among us we put on trial, while incitement in the Palestinian Authority takes place in official media outlets, in the education system, incitement centered on the call to destroy the state of Israel."[38]

However, Palestinian President Mahmoud Abbas and Khdeir's family blamed the murder on the Israeli government for inciting violence against Palestinians by promising "vengeance" for the slain Israeli teens. Abbas claimed, "Israel bears full responsibility for this incident. The Israeli police and security forces must bring those responsible to justice."[39] Abu Khdeir's uncle, Ishaq Abu Khdeir, said, "We demand that the Israeli government find the criminals, and protect the Palestinian population. Everyone in the government is responsible for these crimes, from Netanyahu down."[40]

Secretary of State John Kerry issued a statement saying Washington "condemns in the strongest possible terms the despicable and senseless abduction and murder [of Abu Khdeir.] It is sickening to think of an innocent 17-year-old boy snatched off the streets and his life stolen from him and his family. There are no words to convey adequately our condolences to the Palestinian people. [Those] who undertake acts of vengeance only destabilize an already explosive and emotional situation. We look to both the Government of Israel and the Palestinian Authority to take all necessary steps to prevent acts of violence and bring their perpetrators to justice. The world has too often learned the hard way that violence only leads to more violence and at this tense and dangerous moment, all parties must do everything in their power to protect the innocent and act with reasonableness and restraint, not recrimination and retribution."[41]

Rising Tensions

These events in the West Bank occurred while Hamas and other terrorist organizations intensified their rocket and mortar attacks, targeting Israeli civilians from the Gaza Strip.

During the previous conflict in 2012, rockets from Gaza only reached as far as Tel Aviv (40 miles) and Jerusalem (60 miles), but newly acquired M-302 rockets had allowed for militants in Gaza to strike Hadera, located in-between Haifa and Tel Aviv, 73 miles north of Gaza.[42] According to the IDF, between June and July 8, "militants fired 250 rockets capable of reaching Israel's largest cities and populations and endangering 3.5 million Israeli lives . . . more than 70 percent of Israelis live within range of Hamas' rockets."[43] This put immense pressure on Israel to neutralize the threat.

Israel adamantly wanted to avoid escalation with Hamas in the Gaza Strip, as its aim was to reduce Hamas' military capacity in the West Bank. However, its diplomatic efforts failed to broker an agreement or international intervention. Notably, the Palestinian government was also hoping to avoid an escalation. When asked if he wanted Hamas to expand the strikes to Tel Aviv, Ali Salam, a Palestinian government official, responded: "We will all eat manure if we do that. I want the Egyptians to mediate and end this issue."[44]

During this time, the IDF uncovered Hamas' extensive tunnel system. This vast network of cross-border attack tunnels enabled militants to infiltrate Israel and carry out terrorist attacks. According to IDF Spokesman Lt. Col. Peter Lerner, "Hamas had a plan. A simultaneous, coordinated, surprise attack within Israel. They planned to send 200 terrorists armed to the teeth toward civilian populations. This was going to be a coordinated attack."[45] According to Hirsh Goodman from the Institute for National Security Studies, "This was potentially Hamas' terrorist version of the 1973 Yom Kippur War, when Egypt and Syria launched a joint surprise attack on Israeli forces in Sinai and the Golan Heights. In this case, however, Israel's cities were to be the battlefields and civilians the victims of war. It would not have been an attack to regain territory lost in war, but an indelible reminder that Hamas would never accept Israel's existence."[46]

Warfare in Gaza – General Observations

As tensions escalated in Gaza, civilian life in southern Israel became "unlivable." At the beginning of the government session on July 6, 2014 Prime Minister Netanyahu stated that the IDF had already started carrying out military operations against many targets in Gaza. "Our goal is to restore peace and calm to all Israeli people, and in particular to the people who live in the south. We are conducting our activities in a balanced and cold minded way."[47]

Prime Minister Netanyahu knew that the decision to go to war against Hamas was supported by a solid majority of the Israeli public. According to a survey of The Israel Democracy Institute, throughout July 2014, a vast majority of the Israeli public supported the war: on July 14, 85.3 percent of the Israeli people believed the war was "absolutely justified," while 10.2 percent thought it was "somehow justified." On July 17, 2014 the number of those who believed the war was "absolutely justified," dropped to 71.4 percent, while 20.8 percent believed it was "somehow justified." On July 23, 82 percent believed it was "absolutely justified," while 14.9 percent believed it was "somehow justified." About 50 percent of Israelis believed the Israel used an appropriate level of fire during the fighting. The other 50 percent believed the level of fire Israel employed was too small.[48] As the war already began Prime Minister Netanyahu made the following statement:

> Israel is in the middle of a military campaign designed to bring peace and security to our people. We will not tolerate the launching of missiles against our cities and villages. Our decision to expand the war against Hamas had been undertaken only after all the political efforts to restore peace have failed. We do not want war, but we are committed to ensure the safety of our people. Our acts are directed against Hamas, not against the people of Gaza.[49]

In a meeting with army commanders the day after the operation began, Prime Minister Netanyahu said that Hamas would pay a high price for its aggressive acts against Israel.[50] On July 9, 2014, Prime Minister Netanyahu informed the public that he spoke with various world leaders. He made it clear to them that Israel was determined to protect its citizens. He condemned Hamas for the war crimes it was carrying out. "Its people are hiding their rockets in schools hospitals and mosques knowing that we do not want to kill innocent people. At the same time, Hamas is targeting its missiles against ordinary Israeli people." Netanyahu made the point that he highly appreciated the fact that all of the world leaders with whom he spoke made it clear that they affirmed Israel's right to defend itself against Hamas.[51]

In a joint conference with the Defense Minister Moshe Ya'alon, Prime Minister Netanyahu stated that Israel had inflicted a heavy blow on Hamas, and that it would continue to do so. Yaalon said that he was satisfied with the way the operation had been handled so far. He also cautioned against expectations of an immediate end to the war.[52] The IDF spokesman stated on July 12 that there are clear signs that Hamas was under stress. Hamas leaders knew that the residents of Gaza would pose tough questions to them when the war ended. Israel would exhaust every option at hand to hit Hamas, however; this would be done according to international law.[53]

In the coming days, the Israeli people became aware of the complexity of the fighting in Gaza. As the number of Israeli casualties increased, criticism of the government grew for the "soft hand" with which it was dealing with the Palestinians in Gaza. Many called for the resumption of direct targeting against the leaders of the Hamas, but at this stage the Israeli government seemed to be reluctant to take this action. At a press conference on July 11, 2014, Prime Minister Netanyahu tried to explain the reasons for avoiding the direct targeting of Hamas leaders. He stated that the intensity of the operations carried out against Hamas is very high. "It is double the intensity of Operation Pillar of Defense," he explained. Implicitly, he wished to say that there was no need at this stage to undertake a highly escalatory act. Netanyahu added that there were no immune targets in Gaza for Israel, and that Israel would hit any target that it deemed necessary to hit.[54]

"However," he added, "we should know that the Hamas leaders and officers are hiding within a highly dense population." He meant to say, though he did not say it explicitly, that direct targeting would necessarily cause the deaths of innocent civilians. "Whatever we would say and however we would try to explain or justify these acts, we would stand against harsh criticism of world public opinion. This might eventually limit our freedom of action. So the price we would pay for the killing of a Hamas leader or high-ranking officer would be much higher than the benefit we would get from it."[55]

In any case, the pictures that came out of Gaza's hospitals started to have a negative effect on the legitimacy of Israel's operation in Gaza. From the very beginning, this warfare seemed to be conducted between unequal sides; after all, Israel was seen as a superpower compared to Hamas. Under these circumstances, the sympathy of the public worldwide was given to Hamas rather than to Israel. Furthermore, Hamas had an advantage over Israel in its ability to use the media for the service of its interests. Hamas could come out immediately with statements regarding the responsibility of Israel for various tragic occurrences in Gaza. Israel did not enjoy the same privilege. Its authorities had to check very carefully every statement given to the media in order to ensure its accuracy.

In an attempt to shift public opinion in favor of Israel, Prime Minister Netanyahu gave interviews to leading media network. During these interviews, Netanyahu emphasized the following main points:

1   Our aim is just to restore peace and tranquility to the region and in particular – the southern part of the Negev. Implicitly, he noted that Israel's aim was not to defeat the Hamas. Nor did it seek even to destroy its infrastructure.

2 This goal, Netanyahu said, will be achieved either by military or political means. This was a clear message to the Hamas that Israel is not interested in prolonging the war and is ready to carry out talks to end the conflict.[56]

3 While Israel uses every means possible to refrain from harming civilians both in Israel and in Gaza, Hamas hides its weapons in hospitals, schools, and mosques. It targets its rockets against innocent civilian in Israel, and demonstrates indifference to civilians who might be killed in Gaza.

4 Everyone who wishes to criticize Israel must answer the question of how he or she would like their government to respond if their country was being shelled night and day.

5 The warfare in Gaza only enhances Israel's position to ensure that Iran does not become a nuclear state. Iran is the country that supplies arms to the Hamas, and is directly responsible for the outbreak of hostilities.[57]

In order to demonstrate Israel's willingness to end the war by political means, Netanyahu declared, following a meeting with German Foreign Minister, that Israel accepted the Egyptian suggestion for an immediate ceasefire. Netanyahu stated, "The Egyptian suggestion should lead to the demilitarization of Gaza from missile and tunnels by political means. As it seems now, Hamas would not accept this offer. Therefore we will have all the legitimacy to enlarge the scope of our fighting."[58] And indeed, Israel accepted the Egyptian proposal for a ceasefire, and halted all military activities for six hours. Both the Secretary-General of the United Nations and the Arab League gave their support for the proposal. However, Hamas refused to accept the proposal and continued launching missiles towards Israel.

Hence the perception that Hamas is solely responsible for the continuation of the hostilities and for those who would be killed and wounded, both Israelis and Palestinians alike. In a meeting with the Italian Foreign Minister, Federica Mogherini, Netanyahu stated that if missiles had attacked Italian cities like Rome or Milano as Israeli cities were being attacked, they would have acted as we are acting now. "Hamas," he said, "does not want a political solution to the conflict. It wishes to destroy the State of Israel." And then, for the first time, Netanyahu made it clear that Israel would not only fight against Hamas, but would also defeat them.[59] Obviously, this goal has not yet been fulfilled.

On July 22, the Secretary-General of the United Nations, Ban Ki-moon, visited Israel and met with Prime Minister Netanyahu. During the press conference following the meeting, he tried to reflect a balanced position between Israel and Hamas. However, it was clear that the dominant part of his speech included statements of criticism towards Israel and its policy in Gaza. In his initial remarks, the secretary said he saw the effects of the missiles that Hamas was launching towards populated areas in Israel. "This is indeed shocking," the secretary stated. "The United Nations sharply deplores the launching of missiles against Israel's cities and villages. This has to stop immediately."[60]

"We also deplore the use of civilian bodies for military purposes by Hamas. Israel cannot tolerate the launching of missiles to its territory. All states have a duty to defend their people." He also sent his condolences to the parents of the three Israeli teenagers who had been kidnapped and murdered by Hamas. He said that he was particularly moved by the words of Rachel Fraenkel, whose son was murdered by Hamas. She appealed to the mother of Abu Khdeir, whose son was allegedly murdered by extremists Jews: "I feel your pain," she said. "Too many Israeli and Palestinian parents are losing their children. We must make an effort to bring this suffering to an end."[61]

"In recent days," the Secretary-General added, "I met with many leaders who are involved in this conflict. This is part of an intensive international effort to bring an end to the war. My message to both Israelis and Palestinians: stop fighting, start talking. Try to discuss the roots of the conflict so that in a year from now, we will not find ourselves in another military confrontation." Here he clearly expressed views that put some responsibility for the confrontation on Israel. "There has to be mutual recognition between the fighting parties," he stated. "We must deal with the effects of the occupation on the Palestinians so that people will not wish to express their frustrations by violent means. Military operations will not enhance Israel's security and stability in the long run. Israel is a strong democracy," the Secretary added: "I urge you to demonstrate restraint in your military activities. I know that many Israelis are feeling desperate and threatened by the situation. But there is no alternative to one united Palestinian government and the two-state solution. Israelis and Palestinians have to recognize the fact that both of them share a common future. The United Nations will do everything in its power to bring an end to the present conflict."[62]

In his speech, Prime Minister Netanyahu reiterated Israel's well-known position. "Israel," he stated, "is committed to bring an end to the threats and violence posed against its people day and night for a long period of time. We did not want this escalation. We had no interest in it. We accepted the Egyptian proposal for a ceasefire – a proposal that had been supported by the United Nations, the United States, the Arab League, and Europe. We have also accepted the ceasefire proposed by the Red Cross twice. Hamas rejected it. The international community must demand from Hamas explanations for its positions that only prolong the warfare and for the fact that it is using civilian bodies [civilians] for military purposes. Israel, as opposed to Hamas, is taking every possible measure to ensure that innocent civilians will not be hurt – neither in Israel nor in Gaza." Netanyahu added, "As you may recall, the international community exerted heavy pressure on us to supply cement to Gaza. We sent many tons of cement."[63]

"However, instead of using this cement to build schools and hospitals they used it in order to build tunnels through which their fighters will enter into Israel villages and kill innocent civilians. We have also established a hospital nearby the border in order to help wounded people in Gaza. However, Hamas does not allow its people to use it." At the end of his speech, Prime Minister Netanyahu praised the resilience of the Israeli people. "This is a strategic asset for the State of Israel. Hamas believed that they would break our morale, but they face another disappointment."[64]

As the ground operation in Gaza began, the number of casualties grew. This eroded morale within the Israeli public. The sensitivity for human life in Israel is extremely high, and news about fallen soldiers has a significant effect on the public mood. Besides, it seemed that the public could not understand why an extremely powerful state such as Israel with the most technologically advanced military in the world could not quickly defeat a relatively small terror organization equipped with relatively retrograde arms. The majority of the public could not understand why in a situation of war, Israel was not utilizing all of the powers that it possessed in order to defeat its enemy.[65]

Under these circumstances, Prime Minster Netanyahu felt there was a need for him to make a speech to the nation, and on July 20, 2014, he delivered that speech. At the very beginning, he expressed his deep sorrow and that of the whole nation for the families of the fallen soldiers. He also said that he was praying for the immediate recovery of the wounded soldiers. "The war in which these soldiers died or were wounded is the most justified war Israel has ever fought," stressed Prime Minister Netanyahu. "In these difficult days," he continued, "We have to keep our unity. This is the root of our strength, and we all pray for the safety of our soldiers." "Israel," Prime Minister Netanyahu said, "did not want this war. It was imposed on us. But we will not stop until we achieve the goals we set for us to achieve." Here, he gives his version of the goal of the war: restoring peace to the people of Israel for a long period of time and inflicting a severe blow to the Hamas infrastructure. It was clear that Netanyahu assessed that Israel would not be able to establish an "eternal" deterrence against Hamas. Therefore he limited the scope of his goals to the creation of deterrence "for a long period of time," for which, obviously, he did not define its exact meaning.[66]

"Israel," Prime Minister Netanyahu said, "has already inflicted very severe damage to Hamas' abilities." He mainly alerted to the destruction of many underground tunnels that Hamas built in order to infiltrate inside Israel and attack one of the kibbutzim or villages near Gaza. "During the days that preceded the beginning of our ground operation, we took all necessary measures in order to gain the support of the international community in our campaign. We have agreed to a proposal of ceasefire offered by Egypt. We have also agreed to two proposals for humanitarian ceasefire offered by the United Nations and the Red Cross. We have also conducted intensive dialogue with world leaders." [67]

The fact Hamas rejected several proposals for a ceasefire since the beginning of the war has certainly enhanced the legitimacy of the Israeli operation in Gaza. This was reflected in statements made by leaders who visited Israel during the war. On July 24, the British Foreign Minister Philip Hammond came to meet Prime Minister Netanyahu. In a joint press conference, Netanyahu made a comparison between the missiles launched against Israel from Gaza with the air strikes carried out against Britain by the Nazis. He stated firmly that Hamas was carrying out war crimes both against Israel and against its own people. Finally, Netanyahu praised British Airlines for showing courage in its decision to fly to Israel during the war, notwithstanding the fact that many other air companies chose to cancel their flights to Israel during this period of time.[68]

The British foreign minister made it clear that his government held Hamas responsible for the current military confrontation. "Hamas decided to launch hundreds of rockets towards Israel, abrogating the rules of international law." "Israel," he said, "has full right to defend its people. However, Britain is much concerned about the growing number of casualties of innocent citizens and we urge the parties to reach an agreement on ceasefire as soon as possible. Britain appreciates the fact that Israel accepted the ceasefire and disappointed by the fact that the Hamas rejected it. However, Britain is determined to do everything in its power to bring an end to the confrontation."[69]

Under these circumstances it was only natural that Prime Minister Netanyahu would use this opportunity in order to increase international pressure on Hamas. In a speech delivered on July 28, Netanyahu called upon the international community to stop the transfer of money to Gaza until it was clear that the money was being used for civilian purposes. Until now, the money and the cement that was given to Gaza were directed to military goals; foremost among these goals was the buildup of tunnels directed to kill innocent Israeli people. "This situation should not continue any more," Netanyahu said. He also made it clear to the Israeli public that he did not foresee an immediate end to the war. "We are facing a brutal and cruel enemy that is indifferent to the lives of its own people."[70]

The Chairman of Armed Services Committee in the United States Congress, Howard Buck McKeon, gave the most apparent support for Israel's campaign. Following a meeting with Prime Minister Netanyahu a few days after the end of war, McKeon made it clear that Israel and the United States, among all civilized states, are facing common enemies – radical Arab terrorists: "They execute people, they target innocent people, and they show indifference to the lives of their own people." In his view, there was no difference between the cruel acts carried out by ISIS and those carried out by Hamas. "Israel and the United States should join forces in order to defeat these evil organizations." Mr. McKeon expressed the gratitude of the American people for Prime Minister Netanyahu's determination to "stand against the enemies of humanity." "The way Israel is fighting against Hamas, and in particular its intensive efforts to avoid killing of civilians, both Israelis and Palestinians, is a model for all civilized states. We are very proud to cooperate with Israel in the buildup of technologies like the Iron Dome, which helps to save human life."[71]

A day after a ceasefire was agreed and the war ended, Netanyahu and Defense Minister Ya'alon carried out a press conference in which they surveyed the nature of the war and Israel's achievements: Netanyahu stated that the war was a big military and political success for Israel. "Hamas has suffered a severe blow in that we accepted none of its conditions for a ceasefire. Hamas demanded a seaport; an airport; the release of Hamas prisoners from Israeli jails; the transfer of money from Israel via the mediation of Qatar and Turkey; Israel opposed all these demands. Nevertheless Hamas agreed to a ceasefire." During the war, Netanyahu said, that he enjoyed a tight cooperation with the Defense Minister and the Chief of Staff. "During the war, we killed more than one thousand Hamas fighters, including many high-ranking officers. We succeeded in preventing many efforts by Hamas to carry out highly dangerous operations against Israel. The Iron Dome has neutralized almost 90 percent of the rockets launched against Israel. This is the most severe blow the Hamas has suffered since its establishment."[72]

Hamas also suffered severe political damage, as it became isolated both in the region and within the international community. "We have been given a freedom of action for 50 days – a long period of time. This is not something to be taken for granted. We have managed to convince the international community that Hamas, ISIS, and Al-Qaeda are quite similar organizations. During the war, we carried out an intensive dialogue with moderate Arab states. This certainly creates a basis for closer strategic cooperation between us. We have created deterrence against Hamas, and it is likely that we will enjoy a long, peaceful period of time in Gaza. Hamas believed that they would be able to carry out a war of attrition against us. They failed. Hamas was also surprised by the resilience of the Israeli people, and the solidarity within the Israeli society."[73]

As the war in Gaza came to an end, Hamas leader Khaled Mashal, staying in Doha (Qatar), far away from the fighting, gave his viewpoint on war. He praised the fighters of Hamas and the people of Gaza in general, who managed to "stand with great courage against Israeli aggression." Israel, he told his audience, "carried out a holocaust much more severe than the one Hitler had carried out." Nevertheless, Hamas eventually managed to create a sort of balance of terror with Israel. The best proof is the fact that during the war almost five million Israelis stayed in shelters. "The Palestinian resistance," he promised, "would continue notwithstanding the ceasefire."[74]

Justification of the Operation

When evaluating the decision making of the Israeli leadership at the onset of Operation Protective Edge, it is important to note Israel's justification for action in a legal context. When assessing the legality and legitimacy of the operation, Pnina Sharvit Baruch of the Institute for National Security Studies found that from a legal standpoint, "there is no doubt that Israel has the right to use military force in the Gaza Strip to prevent attacks from there aimed at Israel," citing that because Operation Protective Edge was part of an existing protracted armed conflict between Israel and Hamas, Israel did not need to rely on the right to self-defense to take action, as this is only relevant at the beginning of an armed conflict.[75] Baruch continues, "Even if there were need to establish the claim of self-defense, Israel could definitely do so, because Hamas clearly engaged in armed attacks against Israel," adding that, "Israel tried to avoid using force, by offering 'calm for calm,' and by agreeing to a ceasefire." She noted that, "In both cases, it was Hamas that chose to continue the offensive, further supporting the legal justification for Israel's use of force."[76] In determining that the use of force was appropriate given the situation, the next step is to evaluate the nature of its use.

An audit conducted by the State Comptroller of Israel found, "From the minutes of the cabinet discussions which took place between the decision to embark on Operation Protective Edge and its conclusion, as well as from the statements made by the cabinet ministers at the time and other senior officials, it was clear that both the political echelon and the senior military echelon explicitly considered the limitations and rules set forth in international law with regard to the conduct of the fighting in Gaza, and the Prime Minister gave explicit instruction to refrain from harming uninvolved civilians. The minutes also indicate that both the political and the senior military echelons took into account, as part of the conduct of the hostilities in Gaza, the issue of humanitarian assistance to the residents of Gaza."[77] The report also found that the Attorney General in coordination with the Military Attorney General (MAG) continuously provided to the political and military echelons legal advice on compliance with the rules of international law.[78]

Israel officially claimed that throughout Operation Protective Edge, Israel accepted and respected numerous ceasefires, while Hamas violated eleven ceasefire agreements and UN-sanctioned humanitarian windows by continuing to target Israeli civilians and forces.[79] Hamas' rejection of these ceasefires prolonged the conflict. For example, if Hamas had accepted and respected the July 15 Egyptian-brokered ceasefire, Israel would not have launched the ground invasion into Gaza.[80] Additionally, had Hamas accepted the July 15 ceasefire, which featured roughly the same terms that Hamas would ultimately accept on August 26, it is estimated that roughly 90 percent of the casualties incurred during the 2014 Gaza War could have been avoided.[81]

The IDF's code of ethics, "The Spirit of the IDF," states: "The soldier will only use his weapon to carry out the mission, but only to the extent necessary, and will preserve his humanity even in combat. He will not use his weapon and his power to harm noncombatants and captives, and will do everything in his power to prevent harm to their lives, body, dignity and property."[82]

Under the laws of the State of Israel and International Humanitarian Law (IHL; also known as the "Law of Armed Conflict"), the IDF is obliged to ensure that "collateral damage" in military operations is proportionate to the potential military benefit. The number of civilian casualties does not, in and of themselves, necessarily imply that the military force used was disproportionate. Under the Law of Armed Conflict, proportionality depends on the amount of force necessary to repel attacks and to eliminate the ongoing threat.[83]

While civilian casualties are always tragic, in some cases they are inevitable. During Operation Protective Edge, the IDF engaged an enemy that went out of its way to endanger its civilian populace; by contrast, Israel went above and beyond in its attempts to limit civilian casualties while still protecting its own citizens.

Israel has a highly developed state apparatus to ensure that its armed forces obey IHL. The Military Advocate General's Court and the Supreme Court of Israel monitor, examine, and investigate the IDF's adherence to IHL. These institutions comply with international standards. The "Teerkel Commission," established in 2012 to review Israel's mechanisms for monitoring military compliance with IHL, found that the "Israeli investigation mechanism . . . is consistent with Israel's international legal obligations." The Military Advocate General's Corps of the IDF provides legal advice and training on IHL to IDF commanders and investigates suspected violations. MAG is not subject to the IDF chain of command and retains independent discretion. Following the Teerkel Commission's recommendations, the MAG activated a fact-finding assessment mechanism during Operation Protective Edge to expedite investigations. While an audit conducted by the State Comptroller of Israel uncovered some flaws with respects to the efficiency and expediency of the fact-finding assessment mechanism (FFA) during and after "Operation Protective Edge," it was concluded that the mechanism did in fact work in good faith.[84]

The IDF is also subject to judicial oversight by the Supreme Court of Israel, which has emphasized that respect for the rule of law and human rights are essential for Israel's national security and image, stating, "This is the destiny of a democracy –it does not see all means as acceptable, and the ways of its enemies are not always open before it. A democracy must sometimes fight with one hand tied behind its back. Even so, a democracy has the upper hand. The rule of law and the liberty of an individual constitute important components in its understanding of security. At the end of the day, they strengthen its spirit and this strength allows it to overcome its difficulties."[85] According to the Jerusalem Center for Public Affairs (JCPA), "the depth and breadth of judicial oversight of the military by the Israel Supreme Court has no parallel anywhere in the world."[86] Israel's sophisticated network of institutions properly upholds and ensures compliance.

The "Goldstone Report" further incentivized Israel to ensure its strict compliance with IHL. The Goldstone Report found evidence of potential war crimes and "possibly crimes against humanity" by both Israel and Hamas during Operation Cast Lead. Judge Richard Goldstone, head of the mission, retracted the accusation two years later saying, "The allegations of intentionality by Israel were based on the deaths of and injuries to civilians in situations where our fact-finding mission had no evidence on which to draw any other reasonable conclusion. While the investigations published by the Israeli military and recognized in the UN committee's report have established the validity of some incidents that we investigated in cases involving individual soldiers, they also indicate that civilians were not intentionally targeted as a matter of policy."[87]

Despite Goldstone's retraction, the damage to Israel's international reputation had already been done. According to JCPA, "Ever since

Goldstone, global efforts to delegitimize the State of Israel and its policies have gained unprecedented impetus. In the wake of this and other experiences, no one in the Israeli military establishment fails to comprehend that civilian casualties play into the hands of the enemy."[88] During Operation Protective Edge, therefore, Israel was highly motivated to ensure proper compliance with IHL in order to rebuild its international reputation.

Well-equipped and highly motivated, Israel then cemented its obligation to limit civilian casualties through several operational measures. Israel gathered real-time intelligence information, used high-precision munitions, carefully timed attacks, and provided advance warning to civilians in order to improve the military's capabilities to strike specific military targets while reducing the risk of damage to civilian life. It should be noted that over the course of Operation Protective Edge, Hamas officials repeatedly directed civilians (sometimes by force) to ignore these warnings in an attempt to shield military assets from IDF attack.[89]

The *Wall Street Journal* reported such an incident when seven Palestinians were killed after running into a house in Khan Younis despite warnings from the IDF that it was about to be bombed. According to the report, surviving family members said they believed that if they stayed as human shields, they could stop the attack, and dozens of men rushed to the roof.[90] This is a direct violation of Article 28 of the Fourth Geneva Convention, which clearly states, "The presence of a protected person may not be used to render certain points or areas immune from military operations." Nevertheless, Israel went to extraordinary measures to limit civilian casualties, sometimes at its own disadvantage. For example, prior to its operations to destroy cross-border assault tunnels in the Gazan neighborhood of Shuja'iyeh, the IDF repeatedly warned residents to evacuate over the course of three days, and even postponed its operation another 24 hours in order to allow for as many residents as possible to leave. This extraordinary effort by Israel to minimize civilian casualties cost the IDF the operational element of surprise, and allowed for Hamas militants to fortify themselves for combat.[91]

Under the Law of Armed Conflict, members of organized armed groups, including but not limited to the military wing of Hamas, may be attacked at any time unless they become injured or serve a function that entitles special protection (such as medical personnel).[92] The measures mentioned earlier in this chapter are implemented by the IDF to limit collateral damage and to ensure that their targets are legitimate in the context of international law. However, in a dense urban environment like the Gaza Strip, targeting such individuals comes with an increased risk of collateral damage. Hamas fully understood and exploited this fact as part of their strategy throughout Operation Protective Edge. Despite this, the majority of Israel's 6,000+ airstrikes during Operation Protective Edge resulted in no civilian fatalities.[93]

Further actions taken by Israel to assist Gazan civilians during the conflict include: facilitating the movement of 5,637 truckloads delivering 122,757 tons of aid to the Gaza Strip,[94] facilitating the movement of 6,000–7,000 people in and out of the Gaza Strip,[95] the provision of medical care and evacuations, and the establishment of an Infrastructure Coordination Centre to identify needs and conduct repairs with regards to fuel, electricity, water and sewage, and communications within the Gaza Strip.[96] All of these actions were taken despite attempts by Hamas to hinder Israel's humanitarian efforts throughout the 2014 conflict, including the repeated shelling of the Erez and Kerem Shalom crossings, which periodically delayed the transfer of persons and goods.[97]

Specific mention should also be given to measures taken by Israel to minimize Israeli civilian casualties, namely civil defense initiatives such as the construction and use of bomb shelters, warning sirens, and perhaps most notably the Iron Dome missile defense system. While the exact effectiveness of Israel's missile defense system is still up for debate, it has nevertheless had an impact, as Emily Landau and Azriel Bermant of the INSS note, "Missile defenses do create conditions for enhanced freedom of action for decision makers – defense systems ensure that they have time, and are not compelled to resort automatically to pre-emption and retaliation." Landau and Bermant concluded, "It is not just about protecting the Israeli public, but also about enhancing stability and de-escalation efforts."[98]

Hamas Maximizes Civilian Casualties

While Israel was highly selective in its use of force and sought to uphold IHL standards to limit civilian casualties in Gaza, Hamas deliberately exposed its civilians to the dangers of the conflict in order to advance their political agenda. During Operation Protective Edge, Hamas allegedly used civilian infrastructure as cover for their military activity and inflated civilian death tolls in order to use high civilian casualty counts as political leverage. For Hamas, high civilian death tolls bolster their claim that Israel uses "excessive" and "disproportionate" force. This, in turn, damages Israel's international reputation while leveraging the Palestinian cause.

Thane Rosenbaum, Senior Fellow at New York University's Law School and contributor to the *Wall Street Journal*, explains that, "Winning the PR war is the best Hamas can hope to achieve . . . each time the levers of relative loss are reported, world public opinion will turn against the Jewish state and box Israel into an even tighter corner of the Middle

East."[99] This goes to show how modern democratic militaries must increasingly confront "law fare," that is, the use of law as a substitute for traditional military means to achieve war-fighting objectives, when facing unconventional adversaries, particularly in urban environments.[100]

Hamas' preference for urban combat is well documented in training and doctrinal materials seized by the IDF over the course of Operation Protective Edge. Documents such as "Characteristics of Warfare in Populated Areas," and "A Chapter in Urban Combat," emphasize the operational difficulties that urban environments pose to the IDF, and the advantages that it gives to Hamas and other militant groups. Mention is specifically given to the challenges that the presence of civilians presents to the IDF, namely, difficulties in utilizing full firepower, difficulties in controlling the civilians during and after operations, and how damage to civilian life and property strengthens hatred towards the IDF while increasing support for Hamas.[101]

Throughout the duration of the 2014 conflict, Hamas unnecessarily exposed its civilians to the worst of fighting time and time again. This was a deliberate and integral part of Hamas' strategy that served to impair the IDF's ability to operate and to foment anti-Israel sentiment. According to Maj. Gen. (Res.) Shlomo Turjeman of the IDF, "Hamas came to realize that fighting within highly populated areas offered a win-win situation, forcing the IDF into one of two bad choices." Turjeman continues, "The Israeli forces could decide not to attack in areas where civilian casualties would be too high, exposing millions of Israeli civilians to rocket fire and allowing Hamas to protect itself. Alternatively, the IDF could attack Hamas in these areas, which would result in civilian casualties, tarnishing the image of the IDF internationally and stoking a perception that it causes damage to women and children.[102] Any failure to acknowledge this reality when evaluating the 2014 Gaza War plays directly into the Hamas narrative, shielding them from culpability while enabling the continued use of tactics specifically designed to maximize civilian casualties.

During Operation Protective Edge, Hamas reportedly used mosques, hospitals, ambulances, schools, and UN facilities for military activity. According to Israel's Ministry of Foreign Affairs, "[terrorists] also built tunnels, launching sites and weapons-storage facilities near playgrounds in Gaza.[103] In addition, United Nations Relief and Works Agency for Palestinian Refugees (UNRWA) publically stated that it had discovered weapons caches in three different UN schools over the course of two weeks.[104]

With regards to Hamas' use of UN facilities for military purposes, Israel's Ministry of Foreign Affairs stated: "In conducting military operations within UN buildings, Hamas and other terrorist organizations frequently caused them to lose the protections afforded to civilian objects under customary international law, and rendered them legitimate military targets. Operating near these facilities further endangered these structures by drawing IDF counter-strikes towards their vicinity, increasing their susceptibility to incidental damage." The Law of Armed Conflict states that military objectives by nature, including seemingly civilian infrastructure that is being used for military purposes, are lawful objects for attack.[105]

In an August 2014 report, the IDF claimed that Hamas launched over 1,600 rockets from civilian sites and concluded: "Hamas' tactics flagrantly violate international law and the most basic of moral precepts. Given these tactics, the ultimate responsibility for the damage done to civilians as well as the civilian infrastructure of Gaza lies with Hamas."[106]

Some tried to argue that the Gaza Strip is too small and densely populated for Hamas to conduct military operations that do not expose civilians to danger. Even if so, Hamas fails to take any action to limit such casualties. According to Jeffrey White, Defense Fellow at the Washington Institute, Hamas could limit civilian casualties by:

1 Moving weapons and fighters away from built-up areas.
2 Evacuating civilians from defended localities.
3 Stop using civilian buildings for military purposes.
4 Using the military tunnel system as civilian shelters.
5 Putting fighters in uniform.
6 Stop using ambulances and medical facilities for military purposes.[107]

White argued, "These are basic measures that ask nothing more of Hamas than to acknowledge the risk its military operations pose to its own population, and accept some increase in casualties among its fighters instead."[108]

However, White continues, Hamas is unlikely to implement any of the aforementioned solutions because "using civilians as human shields can be an effective military strategy, and there is no real political or military incentive for Hamas to act otherwise – in fact, these proposals would be to the group's political and military disadvantage in the current climate. As long as the world sees Israel as the primary mechanism of civilian casualties, and as long as many Gazan civilians continue to be more concerned with 'resistance' than their lives, Hamas has no reason to change its way of [conducting] war."[109]

Taking this into account, it is misleading to place the blame for Gazan civilian casualties solely on Israel's shoulders. Hamas bears the responsibility to shield its civilians from conflict. Instead, it unnecessarily exposes them to conflict by acting "more like a guerilla group fighting an insurgency than a government responsible for the safety of its citizenry."[110] And herein lies the issue: The United States' criticism of Israel's military operations was misguided and shortsighted as it failed to appreciate the larger political context.

Hamas also augments civilian casualty numbers to form the basis of accusations that Israel is "excessive" and "disproportionate" in its use of force, which, in turn, incites international condemnation of Israel and bolsters the Palestinian case. Hamas called on information officials to claim every casualty as an "innocent civilian" regardless of the facts.[111] On July 17, the Hamas Interior Ministry posted a video with the following instructions: "Anyone killed or martyred is to be called a civilian from Gaza or Palestine, before we talk about his status in jihad or his military rank. Don't forget to always add 'innocent civilian' or 'innocent citizen' in your description of those killed in Israeli attacks in Gaza."[112]

In an August 10 report, the official Hamas television channel in Gaza, al-Aqsa TV, claimed, "Even the jihad fighters in the battleground are actually Palestinian civilians fulfilling their religious and national duty. This is why we say . . . a 'civilian car,' 'a civilian target,' and so on, since we have no regular army and no real military targets, as the occupation is trying to claim in its propaganda."[113] It is important to note that during the 2014 conflict, militants from Hamas and other terrorist organizations frequently disguised themselves as civilians when carrying out attacks.[114] This tactic was particularly frustrating for the IDF, as it impeded the IDF's ability to identify combatants, while also allowing for Hamas to inflate the number of "civilian" casualties.

Palestinian fatality statistics reflected this skew, as disproportionately large percentages of "civilian" casualties in Gaza were men of fighting age.[115] According to a *Time* report, "The demographic analysis of the fatalities in the Gaza conflict has limitations. It can't identify who is or isn't a combatant. But the spike in fatalities among males starting in their late teens and peaking in their early to mid-twenties, and the divergence of the pattern of fatalities from the demographic pattern of the population, raises considerable doubt about claims that as many as 75 percent or more of the fatalities are non-combatants."[116] The report concludes that, in light of the calls to inflate civilian death tolls, "journalists have a responsibility to convey this uncertainty to their audiences and not present figures provided by Hamas and Hamas-affiliated sources as unqualified fact."[117]

Anthony Reuben, head of statistics at BBC News, warned of this. Citing a *New York Times* analysis, Reuben found that "the population most likely to be militants, men aged 20 to 29, is also the most over-represented in the death toll. They are 9 percent of Gaza's 1.7 million residents, but 34 percent of those killed whose ages were provided."[118] Meanwhile, an analysis by IDF experts found that as of April 2015, at least 44 percent of Palestinian fatalities from the 2014 conflict had been identified as militants affiliated with Hamas or other extremist organizations operating in the Gaza Strip.[119]

Nevertheless, Hamas' "death strategy" worked. In its criticism of Israel for failing to limit harm to civilians in Gaza, the Obama Administration failed to appreciate the greater political context regarding Hamas' use of human shields and inflation of civilian casualty statistics. This belittled Israel's attempts to minimize civilian casualties, placed an implicit blame on Israel for Gazan suffering, and limited Israel's freedom of military action.

Military Phases

Operation Protective Edge officially began on July 8 and lasted through August 26, and can be divided into three phases: Aerial campaign (July 8–16), ground offensive (July 17–August 4), and the withdrawal and eventual ceasefire (August 5–26).

Planning for the operation was an imperfect process, as the former head of Israeli Military Intelligence, Major General Amos Yadlin, noted, "There was no specific intelligence indication or strategic warning about the approaching conflict, as demonstrated by cuts in the 2013 defense budget, the reduction in reserve soldier training, and the cessation of IAF training flights."[120] However, despite the lack of specific intelligence, many in the Israeli General Staff and Southern Command believed that given the nature of the relationship between Israel and Hamas, another conflict was only a matter of time.

In 2014, the IDF estimated that Hamas had between 25,000 and 30,000 armed fighters, the majority of which were broken into six brigades stationed throughout Gaza, consisting of anywhere between 2,500 and 3,500 men equipped with mortar teams, anti-tank units, snipers, and infantry.[121] Before the outbreak of fighting, the IDF had to consider both what Hamas capabilities it would have to destroy to bring an end to the fighting, and what Hamas' response would be. These analyses lead to the development of three plans of varying scale (small, medium, and large). However, none of them would match what actually occurred in the 2014 conflict. This is likely due to the IDF's failure to foresee the significant role that tunnel warfare would play in this conflict, as well as unrealistic expectations about the duration of the war in the upper echelons of the military command.[122]

Most of the operational aspects of the campaign were assumed by the

IDF's Southern Command, which controlled three divisions: the 36th Armor Division, the 162nd Division, and the 643rd Gaza Territorial Division, each of which controlled multiple brigade task forces.[123] The IDF also included several special reconnaissance units, and while Israeli Special Forces were utilized during the operation, their specific missions remain classified.[124] In total, the IDF had a 3:1 numerical advantage over Hamas, and enjoyed air, fire, and intelligence superiority, prompting Bar-Ilan University Professor Eitan Shamir to note, "The challenge was, therefore, not whether it accomplished its mission, but rather when and at what cost in human life."[125] Israel would mobilize close to 86,000 reservists for Operation Protective Edge.[126]

As ground units moved into position, the Israeli Air Force was already conducting airstrikes in Gaza. The first phase of Operation Edge saw the IAF conduct over 1,700 airstrikes, targeting weapons storage and manufacturing facilities, rocket launch sites, command and control centers, training and military compounds, military administration facilities, and individual senior Hamas commanders in an attempt to degrade Hamas military capabilities and to end the attacks on Israel.[127] Simultaneously, the Israeli government was making efforts to de-escalate the conflict, as shown by the acceptance of an Egyptian-brokered ceasefire proposal on July 15, a proposal that Hamas rejected.[128]

Despite Israel enjoying complete air superiority, Hamas' tactics would prove to be incredibly frustrating for the IAF. Hamas operatives were well hidden, posing significant challenges to targeting. In addition to this, defensive tunnels constructed throughout the Gaza Strip provided cover to Hamas fighters. Perhaps most frustrating of all, Hamas was and remains deeply imbedded in densely populated urban areas, making the issue of collateral damage paramount in assessing any potential strike.

> There was no doubt that Hamas went through a [process of] thorough learning and force building, because they went under- ground. Not just the offensive tunnels, the whole system under the ground. So when we came with our mighty air force, they were all underground . . . We only comprehended the full system after the operation, and now understood how irrelevant the Air Force really was . . . [The] IAF attacked [say] 1,000 targets, which belonged in the command control family. Did we have any impact on Hamas command and control? No. We attacked thousands of launch sites and had no real impact. So that's why they were so well organized. We came with the Air Force, but those capabilities were irrelevant . . . So for Hamas it was more than just attack tunnels; it was the whole complicated system underground, because they know how good the capabilities of the Air Force are.[129]

In addition to preplanned strikes, the Israeli Air Force and Navy were constantly on the lookout for new targets, while also making serious efforts to minimalize civilian casualties, which Hamas had already been leveraging to delegitimize Israel internationally. The IDF had a process in place to vet and approve targets in a manner that prioritized the minimization of civilian casualties, accepting the disadvantage of increased time between target identification and the execution of an airstrike. Avoiding collateral damage was a primary restriction, and as a result "the IAF did not even utilize a small portion of its capabilities."[130]

On July 17, 2014, thirteen heavily armed Hamas militants emerged from a tunnel near Kibbutz Sufa in Israel. While the IDF was able to detect and kill the Hamas militants, the prospect of terrorists emerging from underground sparked panic in those living in close proximity to the Gaza Strip. While it is up for debate, it has been argued that it was only after the Sufa incident that the Israeli cabinet convened and approved a ground attack on the tunnels.[131] Regardless, these tunnels posed a much more serious threat as opposed to other methods attempted by Hamas (by air and sea), and to eliminate this threat Operation Protective Edge needed to transition from an aerial campaign into a limited ground incursion.

Udi Dekel and Shlomo Brom from the INSS analyzed the advantages and disadvantages of a limited ground incursion in their publication, "The Second Stage of Operation Protective Edge: A Limited Ground Maneuver." Dekel and Shlomo found that in terms of advantages, a successful ground campaign could allow for the IDF to largely neutralize the tunnel threat, take control of a limited area that could be used as a bargaining chip in negotiations for a ceasefire, create a security zone near the border, and enable shorter ranges for operations penetrating deep into the Gaza Strip.[132] The key disadvantage of IDF troops entering Gaza is their susceptibility to a variety of hazards, including but not limited to booby-trapped tunnels, mines, ambushes and anti-tank fire.[133] Nevertheless, as the best way to neutralize the tunnel threat from Gaza, Israel prepared for a ground invasion.

With approximately 900 full-time personnel, Hamas dug three types of tunnels: Offensive tunnels for cross-border assaults into Israel, defensive tunnels to link points in Gaza and maintain communications, and smuggling tunnel's into Egypt which provided Hamas between 40–75 percent of its revenues.[134] In order to destroy these tunnels, the IDF needed to locate and destroy their start points located roughly three kilometers into the Gaza Strip. Notably, 72 percent of the damage done to the Gaza Strip during Operation Protective Edge occurred within this three-kilometer stretch[135] The IDF faced significant challenges in identifying, clearing, and destroying these tunnels, forcing the IDF to experiment with different techniques. Ultimately, the IDF would discover a total of 100 kilometers of tunnels in the Gaza Strip, including 32 cross-border tunnels.[136]

The second phase of the operation began on July 17, 2014, and just two days later the fiercest battle of the war was fought. Located in the heart of Gaza City, Shuja'iya is a densely populated neighborhood with approximately 100,000 civilians.[137] Additionally, it was also a Hamas stronghold, where the IDF approximated that 8 percent of all 1,700 rocket-launches between the start of Operation Protective Edge and July 19 originated from, where Israeli intelligence believed 800–900 Hamas fighters to be located, and where at least six tunnels into Israel originated.[138]

Prior to the battle, from July 16–19, 2014, the IDF dropped 150,000 leaflets, broadcasted over television and radio, and phoned the residents of Shuja'iya warning of the impeding attack, urging them to vacate, but not all complied.[139] Uncertainty about the exact location of Hamas positions and the number of remaining civilians limited the IDF's ability to strike targets via air prior to the attack, and repeated warnings to the areas residents forfeited any element of tactical surprise on the part of the Golani Brigade.[140] The assault commenced on July 19, as the Golani Brigade faced fierce resistance, booby-trapped houses, and an intricate network of tunnels. As the situation on the ground deteriorated, the IDF would turn to firepower, with artillery firing approximate 600 rounds on the Shuja'iya, and the IAF dropping 100, 2,000-pound bombs.[141] The battle ended the next day.

The IDF lost 13 soldiers in the fighting, and Palestinian casualty estimates vary from source to source, but the director of Gaza's Shifa hospital counted 65 killed, including 35 women, children, and elderly, and 288 wounded.[142] Palestinian officials accused Israel of carrying out a "massacre," as the high number of civilian casualties prompted criticism of Israel's military operation from the UN and even the United States. This was not the only controversy of the battle either. Inside Israel, many noted the IDF's tactical error of not having enough Namer armored personnel carriers (APCs), forcing some of the forces to rely on much older M113s, a mistake that would ultimately cost the lives of seven IDF soldiers.[143]

In its defense, the IDF emphasized the repeated warnings issued to Shuja'iya's civilians ahead of time; a move that risked tipping off Hamas militants, putting Israeli soldiers in increased danger, while IDF officials also argued that the areas of most intense destruction overlapped with military targets.[144] With regards to the use of M113s in Gaza, many IDF officers admitted that tactical mistakes were made, but deny any further fundamental issues.[145]

When drawing conclusions with regards to the Battle of Shuja'iya, a report released by the RAND Corporation noted:

> When conventional troops clear irregular forces from a dense urban center and meet determined resistance, the result more often than not is massive destruction and, unfortunately, civilian loss of life. The grim reality is also that the amount of destruction may have altered Hamas's political calculus. According to an Israeli journalist who interviewed Hamas officials during the war, Hamas expected that its ability to inflict significant IDF casualties during the battle would boost their domestic support within Gaza, but the support never materialized. This was because killing Israelis wasn't worth the devastation in Shuja'iya. Around this time, public opinion started to change because they didn't feel they were getting the benefits for tolerating the bombing of Gaza by Israel. The Palestinians eventually had enough.[146]

In contrast to the fierce resistance that the Golani Brigade encountered at Shuja'iya, the 162nd Division and the 401st Armor Brigade faced significantly lighter opposition during their operations in Beit Hanoun and Beit Lahia in northern Gaza, and suffered lighter casualties than their counterparts.[147] The most controversial event to emerge out of the 162nd Division's area of operation was likely the July 24 airstrike targeting a UNRWA school near Beit Hanoun that left 15 Palestinians killed and 200 wounded.[148] It remains a source of controversy today.

The 188th and 7th Armored Brigades fought in the center of the Gaza Strip. On July 19, the 7th Armored Brigade crossed into Gaza looking for cross-border tunnels, while the 188th Armored Brigade was notably called in to reinforce the Golani Brigade during the Battle of Shuja'iya.[149]

The 84th Givati, 35th Paratrooper, and 460th Armored Brigades were tasked with the fighting in southern Gaza, which aside from the Battle of Shuja'iya saw some of the fiercest fighting of the 2014 Gaza War. While fighting militants from both Hamas and Palestinian Islamic Jihad, the 35th Paratrooper Brigade and the 460th Armored Brigade routinely encountered houses rigged to explode as they searched for tunnels in Khan Yunis.[150] The 84th Givati Brigade, which anchored the IDF's southern flank, was involved in two notable incidents that were subject to international controversy.

The first incident occurred at the agricultural village of Khuza'a, located just outside of Khan Yunis, where journalists and NGOs accused IDF soldiers of killing civilians and needlessly imposing siege conditions on the village, although the circumstances surrounding the incident remain unclear. Accounts presented by the IDF paint a different picture, describing an intense fight against a Hamas nerve center for terrorist activity in the area, and reports appear to corroborate this. Interrogation reports from captured Palestinian militants detail how the Khuza'a's early childhood education center was used to facilitate the movement of captured Israeli soldiers, along with how Khuza'a's Al-Taqwa mosque also acted as a Hamas command center.[151] Additionally, following a fire-fight in the mosque on July 29, the IDF discovered weapons and two tunnel entrances, and later released photographs.[152] The IDF also attempted to facilitate the delivery of relief supplies and medical help to Khuza'a, but rubble blocking the route delayed the arrival of the supplies.[153] With regards to the alleged abuse of civilians in Khuza'a, the IDF opened several criminal investigations.[154]

The second incident, which occurred on August 1, in Rafah, came to be known as Black Friday, and involved the controversial "Hannibal directive," which states that IDF forces should do everything in their power to prevent a soldier from being captured, even if it means risking the life of the captured soldier.[155] As a ceasefire brokered by the UN and the United States was set to begin at 8:00 a.m. on August 1, Givati forces moved to isolate a tunnel entrance before the ceasefire went into effect.[156] A clash between IDF forces and Hamas militants followed, with the IDF and Hamas presenting different narratives of the event. According to Hamas, the clash began at 7:00 a.m. before the ceasefire went into effect, while Israel claims that the clash began at 9:00 a.m. after the ceasefire went into effect.[157] Following the clash, IDF Lieutenant Hadar Goldin was suspected captured, prompting the IDF to invoke the Hannibal directive, which led to a significant military response of 40 IAF strikes, 1,000 artillery shells, multiple air-launched bombs and missiles, and the use of bulldozers destroyed a large number of buildings and houses.[158]

As the operations continued through August 3, evidence would eventually lead the IDF to conclude that Lieutenant Goldin succumbed to his wounds, although his body was never recovered.[159] NGOs again accused Israel of using disproportionate force, claiming that anywhere from 29 to over 140 Palestinian civilians were killed, while the IDF defended its actions, claiming it used the necessary force to prevent the loss of one of its soldiers.[160] An extensive internal review of the Hannibal directive by the IDF cleared the Givati Brigade commander of wrongdoing. However, the IDF would end up revoking the Hannibal directive in 2016.[161]

With most of the cross-border tunnels destroyed, Israel withdrew its forces from Gaza on August 3. Two days later, Israel agreed to a proposed 72-hour ceasefire, effectively moving Operation Protective Edge to its final phase. The third and final phase of Operation Protective Edge was characterized by a series of multi-day, coordinated ceasefires with periodic upticks in violence, as Hamas and other Palestinian militants would often violate ceasefires with mortar fire prompting Israeli airstrikes in return.[162] Between August 5 and August 18, Egypt attempted to broker a ceasefire, but it never came to fruition as Hamas and Israel continued to exchange fire. At this point, the 51-day long conflict had taken a toll on both combatants. On August 22, in a sign of internal discord, Hamas executed 18 Palestinians accused of cooperating with Israel, and on the Israeli side, popularity for Netanyahu and faith in Israel's success in the war were fading. Finally on August 26, an Egyptian-brokered ceasefire was able to put an end to the fighting.[163]

By the end of Operation Protective Edge, Israel had lost 66 soldiers and 6 civilians, in addition to the economic costs of $55 million in direct damage to private and public infrastructure, and another $443 million in indirect damages thanks to economic disruptions caused by the conflict.[164] The number of Palestinian casualties is debated. The UN estimated that 2,133 Palestinians died, of whom 1,489 were civilians, while Israeli estimates suggest that out of 1,598 Palestinian deaths, 75 percent were combatants.[165] Physical damage to Gaza was immense; as the UN estimated that up to 500,000 people were internally displaced, while approximately 108,000 people had their homes destroyed.[166] In addition to the cessation of hostilities, the Egyptian-brokered ceasefire increased the area that Palestinians were allowed to farm and fish, while issues such as prisoner swaps and reconstruction were shelved for longer-term negotiations.[167] Many observers also noted that the terms of the August 26 ceasefire were remarkably similar to those proposed on July 15.[168]

Reflecting on Operation Protective Edge, Major General Yadlin, the former head of Israeli Military Intelligence and Director of the Institute for National Security Studies, called the campaign "an asymmetric strategic tie," noting that while Hamas suffered enormous blows on the battlefield, its leadership remained intact and likely improved its standing at home, and while Israel did not make any strategic concessions of consequence, it did not improve its situation dramatically either.[169] The Israeli public was not particularly enthusiastic either, with a *Haaretz* poll finding that most (54 percent) respondents believed that neither side won.[170] This is not to say that the Israeli public was never hopeful with regards to the conflict, as a survey conducted by Rafi Smith for INSS found: on July 27–28, in the midst of the ground campaign, 71 percent of the country's Jewish population thought that Israel was winning the war, with only 6 percent believing that Hamas was winning, and 23 percent considering the conflict to be a draw.

However, when a similar survey was conducted on August 6, shortly after the withdrawal of Israeli ground forces, only 51 percent of respondents believed that Israel had won, the war, with 4 percent believing that

Hamas had won, and 45 percent saying that neither side won.[171] In their analysis of the impact of the operation of political and social trends in Israel, Meir Elran, Yehuda Ben Meir, and Gilead Sher of the INSS found that the fluctuation of public opinion in Israel with regards to the conflict was reminiscent of the Second Lebanon War, both being conflicts that lasted much longer than anticipated with less than conclusive results.[172]

Conversely, many IDF officers and outside experts considered Operation Protective Edge to be a small victory for Israel, pointing to the relative quiet that has been observed on Israel's border with Gaza which they attribute to effective deterrence from Operation Protective Edge. Mark Heller (INSS) noted, "While 77.6 percent of Gaza respondents believed that Israel had been 'painfully beaten by Palestinian militants,' 72.5 percent were also worried about another military confrontation with Israel, suggesting that a new Hamas-initiated confrontation might be received with some lack of enthusiasm."[173] Conversely, reports from Gaza from mid-2015 suggest that Palestinians in areas most devastated by Protective Edge are angry at Hamas' political wing for accepting a ceasefire agreement with Israel that offered no meaningful benefits; however, support for the military wing of Hamas, the Al-Qassam Brigades, remains high.[174]

As one senior Israeli defense official summarized: "In a way," he stated, "there is some deterrence between these conflicts. Everyone understands the price. Every day since August 2014 that Hamas hasn't shot rockets and has arrested people who try to do, so that is deterrence. Israel lets trucks go into Gaza every day. Israel is more engaged than anyone else in addressing the humanitarian crisis in Gaza. The Israelis understand that the humanitarian crisis could cause another war. They want to keep the lid on Gaza; this is a strategic calculus."[175]

While deterrence may have been restored for the time being, by no means should it be assumed that this was a decisive battle, or that Israel and Hamas will not go to war again. As long as the current economic and humanitarian situations in Gaza remain dire, the potential for conflict remains. This is understandably a frustrating reality for those in Israel who had hoped that Operation Protective Edge would finally crush Hamas once and for all, but this was never the objective. As Senior Fellow Major General (res.) Yaakov Amidror at the Begin-Sadat Center for Strategic Studies stated, "The Israeli public needs to understand that Israel did not set out to topple Hamas. Dealing Hamas a debilitating blow, eradicating the terror tunnels and rejecting any change in the status quo that defines Israel's relations with Gaza, were the objectives of the operation. Having achieved all this, the operation's result should not be thwarted."[176] That being said, deterrence is not indefinite, and therefore cannot be considered a "solution," to this conflict, but rather a bandage that will periodically need to be replaced.

Palestinian Authority versus Hamas

The collapse of the Palestinian Unity Government and the events surrounding Operation Protective Edge highlight the nature and the extent of the falling out between the Palestinian Authority and Hamas. From the onset of the chain of events that put Operation Protective Edge into motion, Hamas and Fatah expressed deeply different positions with regards to the "resistance."

In the wake of the kidnapping of three Israeli teenagers in the West Bank, Hamas expressed deep satisfaction and defiance, hoping that the kidnapping along with the ensuing Israeli operations in the West Bank would fan the flames of the "resistance." On June 13, 2014 one day after the abduction, Hamas spokesperson Hussam Badran wrote, "We call upon our people in all parts of the West Bank to confront the occupation, whether as part of mass confrontations or privately- initiated resistance [operations] . . . This is an opportunity to widen the circle of confrontation, and restore the West Bank to its natural status as the spearhead of the resistance."[177] Another Hamas official called upon the Palestinian Authority to avoid cooperating with Israel in the search for the teenagers. The resistance to Israel's occupation is a legitimate right of the Palestinian people.[178]

Other Hamas officials took the kidnapping operation as an opportunity to attack Mahmoud Abbas. Yousuf Rizqa remarked, "The negotiation plan of [PA President Mahmoud] Abbas does not provide the wounded Palestinian people with an alternative [to resistance], nor does the international community provide the Palestinians with any hope. That is why the Palestinian places his faith in his gun and his resistance."[179] Another Hamas official stated that the cooperation between the Palestinian Authority and Israel "brings the Palestinian Authority to its end. Its leaders would be boycotted by the Palestinian people." [180]

Statements from Palestinian Authority President Mahmoud Abbas took a decidedly different tone, condemning the act and those responsible for carrying it out. Abbas stated, "At the end of the day, these are human beings, and we wish to protect human life . . . The truth is that whoever carried out this deed wants to destroy us. Therefore, we will take a different approach toward whoever is behind this, because we cannot tolerate such operations."[181]

Abbas also emphasized and supported continued security coordination with Israel, stating, "Our government believes in security coordination with Israel. There are, of course people who rebuke us and blame us for this, but it is in our interest to have security coordination with the Israelis, in order to protect our people. We do not want to return once again to the chaos and destruction that occurred during the Second Intifada. I say this loud and clear. We shall not return to an Intifada that will destroy us. That is why we have security coordination."[182]

Many members of the Palestinian Authority and Fatah echoed the positions voiced by Mahmoud Abbas, and expressed support for the PA President. On June 19, Jibril Rajoub of the Fatah Central Committee stated, "I oppose kidnapping Israeli and Palestinian civilians . . . kidnapping soldiers is the only solution for freeing the prisoners." On June 20, Palestinian Foreign Minister Riyadh Al-Maliki commented, "Abu Mazen ['Abbas] will prevent a third intifada, despite Israel's military action in the West Bank. Abu Mazen will help find the three missing settlers."[183]

However, that is not to imply that there were no dissenting views in the ranks of Fatah and the Palestinian Liberation Organization (PLO). Palestinian Authority spokesman Adnan Al-Damiri placed the blame for the disappearance of the teenagers squarely on Israel, and on June 14 claimed, "The disappearance of the three settlers is the complete responsibility of Israel, which illegally drew them to the Palestinian nation's territory."[184]

Mahmoud Aloul of the Fatah Central Committee went so far as to propose that the kidnapping of three Israeli teenagers was a hoax, but was soundly critiqued by Sufyan Abu Zaida of Fatah and the PLO, who responded, "This assumption not only shows ignorance regarding the nature of the political regime in Israel and its decision-making apparatuses, but also reflects a disconnect from Palestinian reality. What will proponents of this theory say when the hostages are found? A kidnapping can be faked in tyrannical countries, where the regime controls all three branches of government . . . Israel is democratic with regard to its people and its political regime, and therefore is incapable of putting on such a charade."[185]

From the onset of Operation Protective Edge to its conclusion, Hamas and the Palestinian Authority traded blows over the conflict. Speaking to the Al-Mayadeen TV channel on July 12, Abbas stated, "The most important thing in [this] war is saving lives. The [important] question is not who started it, but how to end it, because Israel has the military advantage, not us . . . the most important thing is to end the hostilities and go back to the understandings of 2012 . . . Israel is attacking us, and we have nothing but the international law to defend ourselves with," adding that, "The martyrs are cannon fodder for the warmongers, and I opposed these warmongers on both sides."[186]

It should be emphasized that Mahmoud Abbas was not only dealing with dissent from Hamas and their supporters, but also from members of his own party. The internal political pressures that Mahmoud Abbas faced should not be understated, with Hussam Khader of Fatah remarking that, "The war in Gaza will bring about the [moral] victory of the resistance and the [establishment of] a Palestinian state in Gaza, and will end Abbas' political career and cause him to leave [the presidency]."[187] Others, such as Jamal Zakout of the PLO and Palestinian Democratic Union, called for Abbas to "Get rid of your advisors and leave for Gaza at once. Visit the wounded and the families of the martyrs, and give your home to a family that has lost its own home... Convey the Gazans' cry to the world, and [call for] a halt to the murderous hostility against the living and for lifting the siege."[188]

The same day of Abbas' remarks, Hamas spokesperson Sami Abu Zuhri responded, "Abbas' statements about the resistance, and his description of the resistance as 'warmongering,' are an offense against the blood of the martyrs and serve positions that are hostile towards our people and the resistance."[189]

On July 22, perhaps sensing growing criticism within his own camp, Abbas gave a major speech at a meeting of the Palestinian leadership in Ramallah, notably with a much more hostile tone towards Israel, in addition to adopting Hamas' narrative and conditions for a ceasefire. "The time has come for everyone to raise their voices and tell the truth, clearly and powerfully, in the face of the Israeli killing and destruction," Abbas exclaimed. "The oppressing occupation forces have crossed every line and [have broken] all the laws. They have deviated from all standards of human and international morality in their ferocity and barbarism."[190] Abbas' speech was a complete turnaround from previous remarks, which had focused on the culpability of Hamas. Abbas also took this time to push for unity, again, perhaps sensing his own vulnerability. "We stress to our people that we adhere to national unity, to ending the schism, and to the national unity government."[191] Statements later released by the Palestinian leadership echoed these remarks.

Despite the apparent pivot towards national unity by the Palestinian Authority, there was little reason to expect that it would actually come to fruition, especially as the war in Gaza continued without an end in sight. During a televised debate on July 23, Fatah Revolutionary Council member Muwaffaq Matar passionately called for an end to the bloodshed. Directing his question at Hamas Political Bureau member Muhammad Nazzal, "How come the torn limbs of children don't move you to make a decision that is not driven by political considerations?"[192]

On the international stage, Secretary of State John Kerry was attempting to broker peace in Gaza, but his actions were coming under intense scrutiny not only by the Israelis, but also by senior Palestinian officials. One Palestinian official anonymously stated that Kerry had tried "to destroy the Egyptian initiative and the Palestinian comments on it (the Abbas Plan) and present a substitute for our initiative . . . Kerry wanted to create a framework that would be an alternative to the Egyptian initiative and to our concept regarding it, in order to please Qatar and Turkey." The Palestinian official continued, "Kerry proposed his initiative after we were very close to a comprehensive agreement guaranteeing the lifting of the siege on Gaza and obtaining all the Palestinian demands . . . An announcement of [this achievement] was ready for publication, but (Hamas political bureau head) Khaled Mashal called a press conference and destroyed the [Abbas] initiative." [193]

As to the motivations behind Kerry's new initiative, the official claimed, "Kerry wanted to take advantage of the war in order to again strengthen the Muslim Brotherhood's regional influence, because the Americans think – and will be proven wrong – that moderate political Islam represented by the Muslim Brotherhood can combat radical Islam."[194] Speaking with regards to the Paris conference, a senior Palestinian official remarked, "This conference was held on the basis of the Kerry initiative, that is, support for and strengthening of the Muslim Brotherhood. We were not invited, and neither was Egypt – but Turkey and Qatar were invited, and that explains everything . . . Abbas is very angry about this profiteering in Palestinian blood and its use in regional power struggles."[195]

Statements released on July 27 by Fatah Central Committee member Azzam Al-Ahmad, as well as from Fatah's information office, express similar sentiments, and a statement from Ahmad Al-Majdalani of the PLO emphasized the importance of the Egyptian initiative stating that there is no other alternative as they hope to distance the Palestinian problem from regional power struggles.[196] Former Palestinian Authority Minister Hassan Asfour went as far as to claim that the United States was actively attempting to get rid of the PLO.[197]

On August 13, Jibril Rajoub of the Fatah Central Committee formally called on Hamas to disengage from the Muslim Brotherhood, signaling a desired rapprochement as he tried to sway Hamas away from political Islam in favor of a more national approach. "The PLO and Mahmoud Abbas have an agenda to establish a state, which vexes the Israeli right, and which is targeted in this war. Therefore, we hope that our brothers in the Palestinian Islamic movement will modify their position so that we may find common ground." Rajoub continued, "They may be Palestinians with an Islamic dimension, but our national identity is what serves as our common denominator. Let go of the Muslim Brotherhood, with all their virtues and their faults. You have seen how political Islam has failed. It has failed."[198]

Operation Protective Edge officially ended on August 26, 2014, but the problems that led to the conflict in the first place have largely remained unresolved, as have the divisions between the Palestinian Authority and Hamas.

3

Policy of the United States throughout Protective Edge

The Balance between "Right to Defend" and "Disproportionate Response"

From Israel's point of view, Operation Protective Edge was first and foremost a defensive action taken as the result of continued aggression from Hamas. However, many governments within the international community, and public opinion in many countries and in the Western media, believed that the issue of responsibility for the war could not be seen in a model of "black or white." As the confrontation began, media outlets quickly became enamored by the "victim narrative" espoused from Hamas and its subjects in Gaza. Images of dead Gazans and bombed out buildings flooded news outlets across the West, prompting outrage towards Israel. As a result, Israel was painted by many networks as bearing some of the responsibility for the outbreak of hostilities, and as a state that employs disproportionate power in its military operations against Hamas. What was completely lost in this coverage was context, and this would have a significant effect on American policy towards Israel.

It has to be acknowledged that Israel unilaterally withdrew from Gaza in 2005 with the hope of achieving at least a lasting armistice with the Palestinian enclave. Nobody believed that a peace agreement with Hamas, a radical organization that openly calls for the destruction of the State of Israel, was a viable option. At the very least, Israelis hoped that they could achieve a period of tranquility: "Gaza," Prime Minister Sharon stated, "cannot be held onto forever. Over one million Palestinians live there, and they double their numbers with every generation. They live in incredibly cramped refugee camps, in poverty and squalor, in hotbeds of ever-increasing hatred, with no hope whatsoever on the horizon. It is out of strength and not weakness that we are taking this step. We tried to reach agreements with the Palestinians, which would move the two peoples towards the path of peace. These were crushed against a wall of hatred and fanaticism. The unilateral Disengagement Plan, which I announced approximately two years ago, is the Israeli answer to this reality. This plan is good for Israel in any future scenario. We are reducing the day-to-day friction and its victims on both sides. The IDF will redeploy on defensive lines behind the security fence. Those who continue to fight us will meet the full force of the IDF and the security forces."[1]

However, peace never came. Instead, rocket fire from Gaza spiked up 500 percent in 2006, and continued to increase in the years that followed.[2] Furthermore, over the course of the 2014 conflict, Israel agreed to eleven ceasefire proposals that Hamas rejected.[3] How is it then that the narrative portrayed by the media came to be so lopsided against Israel? It is because Israel is held as the strong power that fights against a weak power: a Goliath against a David. Naturally, sympathy tends to go towards the weak power. Moreover, Hamas had no hesitation to show the horrors of the war. Pictures of dead and wounded children were repeatedly broadcast on the screens of many people around the world. Nobody could really tell if the pictures were authentic or fake, and nobody could really know who exactly was responsible for these events. Nevertheless, the Palestinians were seen as a victim while Israel was seen as the aggressor. At the end of the conflict, which was the direct result of Hamas aggression, Hamas Prime Minister Ismail Haniyeh announced on *Al-Jazeera*, "Our narrative has gained the upper hand."[4]

The Israeli government headed by Ehud Olmert downplayed the threat that Hamas posed to Israel. In a speech he delivered shortly after Hamas came to power on July 26, 2007, Prime Minister Olmert implicitly stated that Israel has no ability to defend the southern city Sderot from the "flood" of rockets and mortars launched against it by Hamas. He actually called upon the residents of the city to "learn to live" with this threat, knowing that other places where Jews are living are even more dangerous: "We have to tell the residents of Sderot," Prime Minister Olmert stated, "that in the short run we cannot give you the level of personal security which we would have liked to give you. Life in Israel naturally involves some security dangers. He who chooses to live in this country knows that he undertakes to live in a danger. Still, this danger is much smaller than the dangers for Jews who live in other places in the world. We expect all the people of Israel, and in particular those who live near Gaza, to show their resilience, as we all did in the past during much more severe threats."[5]

Another threat that was posed against Israeli residents in the south was the tunnels that Hamas had built into Israeli territory, a project that cost

Hamas approximately $90 million.[6] The danger that these tunnels posed to Israeli communities (many of which are within walking distance to the Gaza border) cannot be overstated. By omitting or downplaying these threats in the international media, and by focusing simply on casualty statistics, Israeli actions appear to be disproportionate. That is the discourse that many covering the conflict took, and it was the discourse that Hamas was relying on for their public relations war against Israel.

From a historical perspective, the Israeli–Arab conflict could be described as David versus Goliath, with tiny Israel playing the part of the former, and the Arab coalitions playing the later. However, as the greater Israeli–Arab conflict began to take a back seat to the Israeli–Palestinian conflict, the roles became reversed in the eyes of many. In the 2014 Gaza War, Israel was painted as a military giant facing off against a tiny Hamas-run enclave. However, a look at the bigger picture shows that Hamas enjoyed significant backing from Iran, Turkey, and Qatar.

Hamas is not simply a small, Palestinian political movement, but a part of a wider Islamist network that stretches well beyond the Gaza Strip, and whose stated goals include the annihilation of the State of Israel. Hamas is also a proxy for much larger powers such as the Islamic Republic of Iran, which share this common goal. Dr. Dore Gold from the Jerusalem Center for Public Affair writes, "The regime in Gaza was not isolated. It received training and weaponry from Iran and Syria. Qatar, which had become a hub for global *jihadist* groups, provided huge amounts of financial assistance. And more recently, Hamas established an operations center in Turkey, which directed attacks in the West Bank."[7] Additionally, some in the West attempted to argue that Hamas had become more moderate, but when Hamas executed 22 Gazans for suspected "collaboration with Israel," even leading Arabic newspapers began to draw comparisons between Hamas ad ISIS.[8]

Perhaps the number one issue affecting perceptions of the 2014 Gaza War was that of civilian casualties. The key problem with regard to the reporting of civilian casualties during the 2014 Gaza conflict is that so many wrongfully assumed that reports from the UN could be trusted. Casualty statistics reported by the UN were, by and large, based on numbers given by the Gaza Ministry of Health, which is effectively run by Hamas.[9] Knowing that it could never topple Israel militarily, Hamas effectively used the issue of civilian casualties to launch a PR war against Israel, and despite the fact that Hamas was using its civilians as human shields for its weapons and personnel, international coverage largely ignored this to pursue further scrutiny of Israel. The numerous and substantial efforts that the Israel Defense Forces took to avoid civilian casualties along with their humanitarian efforts in the midst of the conflict also went largely ignored in international media coverage.

The nature of the media coverage surrounding the 2014 Gaza War and its fixation on the topic of Palestinian casualties certainly influenced American policy towards Israel during the conflict. Throughout Operation Protective Edge, the Obama Administration repeatedly emphasized the need to limit harm to civilians. However, despite its formal commitment to balance and fairness, the administration's statements seemed to sympathize more with Gazan suffering than Israeli suffering. This unbalanced attitude was clearly reflected in Secretary of State Kerry's remarks on July 21, 2014: "We are deeply concerned about the consequences of Israel's appropriate and legitimate effort to defend itself. No country can stand by while rockets are attacking it and tunnels are dug in order to come into your country and assault your people. But always, in any kind of conflict there is a concern about civilians, about children, women, communities that are caught in it. And we are particularly trying to focus on a way to respond to their very significant needs. On behalf of President Obama and the United States, I'm privileged to announce tonight that the United States will immediately provide $47 million in order to provide direct humanitarian assistance of no other kind but humanitarian assistance to try to alleviate some of the immediate humanitarian crisis."[10]

As already noted, from the outset of Operation Protective Edge the Obama Administration took a clear and unequivocal position in support of Israel's right to defend itself; however, this support would always be qualified. In the early stage of the operation, when the main threat to Israel was rocket fire, the administration asked Israel to exercise restraint and limit its activity to aerial responses and refrain from a ground campaign. Once the tunnels surfaced as a concrete threat, the administration again emphasized Israel's right to defend itself, but asked Israel to limit its responses to removal of missile and tunnel threats.

On multiple occasions, the Obama Administration took the trouble to condemn Israel harshly and publicly for significant harm to civilians, particularly near or within UN welfare institutions in Gaza. It seemed that from the administration's point of view, the suffering of civilians in Gaza was a phenomenon, in its own right, that resulted from Israel's military operations. They failed to link to it a greater political context, thereby implicitly faulting the Israelis.

Additionally, when harm to civilians in Gaza was on the agenda, the administration did not even seriously address the admission by UN personnel that Hamas places weapons in UN institutions or the firm demand by members of Congress to investigate the issue.[11] The administration's response to the death of more than ten Palestinians near the UN school in Rafah was especially serious. Officials did not bother to wait for the results of the investigation to confirm whether the IDF was responsible for the event, as is the accepted practice among allies. Jen Psaki used harsh words in relaying the administration's response, stating that, "The United States is appalled by today's disgraceful shelling." According to Psaki, "The coordinates of the school, like all UN facilities in Gaza, have been repeatedly communicated to the Israeli Defense Forces." She added that, "Israel must do more to meet its own standards and avoid civilian casualties."[12]

The wording of the statement left no room for doubt: the administration was not prepared or willing to await the IDF's investigation of the incident, table the matter with a discreet conversation with Israel about such incidents, or accept Israel's claim that it was a tragic error in the use of military force. The US attitude clearly reflected a tendency to see the incident as a deliberate Israeli attack meant to make the residents of Gaza pay a heavy price for the continuation of the fighting. Against this background, the administration apparently sought to further limit Israel's military freedom of action. "The suspicion that militants are operating nearby [civilian sites]," noted the spokeswoman, "does not justify strikes that put at risk the lives of so many innocent civilians."[13]

Again, this reflected a critical failure on the part of the Obama Administration to connect the issue of civilian casualties in Gaza to the larger political context. While Israel was highly selective in its application of military force and sought to uphold its commitment to limiting civilian casualties, leaders in Gaza sought to maximize the exposure of their own civilians to the dangers of the conflict.[14] It is also important to note the words of Charles Krauthammer, whose op-ed in the *Washington Post* rightfully pointed out "the difference between [Israel and Gaza]. We're using missile defense to protect our civilians and they're using civilians to protect their missiles."[15] In failing to drive this point home, the Obama Administration would draw criticism for failing to express its thorough support for Israel, its main ally in the Middle East. Consequently, Secretary of State John Kerry – in at least one instance – took care to reinforce the legitimacy of Israel's actions.

In his autobiography, *Every Day is Extra*, John Kerry recalls the outbreak of the 2014 Gaza War. He acknowledges Hamas tactics that employ the use of civilians as human shields, and emphasized Israel's "right to self-defense."[16]

However, on a formal level, recognition of a state's right to defend itself does not have much significance, since it is the natural and self-evident right of any state to defend itself. This right is also enshrined in Article 51 of the UN charter. However, on the political-public diplomacy level, this affirmation, and the fact that it was emphasized repeatedly by the administration spokesperson, had great significance and was perceived as an expression of American support, even if qualified, for

Israel's military moves. The administration's position was likely influenced by the broad support for Israel in the US Congress and in public opinion during the conflict.

However, at the same time, administration officials repeatedly expressed their deep concern about the severe damage suffered by Gaza's civilian population during the conflict. The hardships sustained by Israel's civilian population were mentioned occasionally, but the administration focused most of its attention on the suffering of "innocent people" in the Gaza Strip. Given the fact that the administration was quite aware that Israel was making every effort to avoid harming the civilian population, the emphatic public comments about the suffering of Gaza residents can be interpreted as an attempt to rein in Israel's military freedom of action in the Gaza Strip. On at least one occasion there was an unmistakable message by President Obama to Israel to ease the military pressure on Gaza, reflected in his assertion that Israel had already caused significant damage to Hamas.

Again, to the Obama Administration, Israel was not entirely free from blame. The Secretary of State was quoted expressing a negative attitude toward Israel and its military campaign, asking, "When is everybody going to come to their senses?"[17] This further placed the actions of Israel and Hamas on the same footing. Finally, it is worthwhile noting the somewhat unconventional, not to mention uncomplimentary remark by US Deputy National Security Advisor for Strategic Communication Ben Rhodes, who conveyed clearly that Israel was not doing all it could to avoid unnecessary civilian fatalities in the Gaza Strip. As evidence, he cited the actions of the US military in Afghanistan, saying, "I think you can always do more. The US military does that in Afghanistan. We go to great lengths. But we believe that in densely populated areas like this, you have to go the extra mile to avoid loss of civilian life."[18]

The Obama Administration's qualified support of Israel during Operation Protective Edge certainly limited Israel's military freedom of action. It also encouraged unfriendly entities to express their attitude towards Israel during the confrontation in a much more hostile manner than they would have had the Obama Administration had shown a greater sense of support towards Israel. This fact did not completely escape the administration's notice. Along with criticism of particular Israeli military actions, the administration expressed its appreciation to Israel's leaders for their efforts to restore calm even at the price of harsh domestic criticism and the appearance of humiliation by Gaza's terrorist organizations.

On July 15, 2014, Kerry made it clear that the escalation of violence entailed great risks and implicitly faulted Hamas with such escalation: "I cannot condemn strongly enough the actions of Hamas in so brazenly firing rockets, in multiple numbers, in the face of a goodwill effort (to secure) a ceasefire. It is important for Hamas not to be provoking and purposefully trying to play politics in order to gain greater followers for its opposition, and use the innocent lives of civilians who may hide in buildings and use as shields and put in danger. That is against the laws of war . . . [There is the] potential of an even greater escalation of violence. We don't want to see that [escalation] – nobody does . . . But Israel has the right to defend itself."[19]

Following the collapse of the ceasefire with the kidnapping and killing carried out by Hamas on August 1, 2014, the administration's attitude toward Israel's military campaign in the Gaza Strip adopted a more positive tone. The US condemned Hamas' actions, terming them an "outrageous violation" of the ceasefire agreement. In remarks to the press on August 1, President Obama said: "I have unequivocally condemned Hamas and the Palestinian factions that were responsible for killing two Israeli soldiers and abducting a third almost minutes after a ceasefire had been announced. And the UN has condemned them as well . . . If [Hamas] is serious about trying to resolve this situation, that soldier needs to be unconditionally released as soon as possible. I have been very clear throughout the crisis that Israel has a right to defend itself. No country can tolerate missiles raining down on its cities and people having to rush to bomb shelters every 20 minutes or half hour. No country can or would tolerate tunnels being dug under their land that can be used to launch terrorist attacks."[20]

Secretary of State Kerry mirrored President Obama's remarks: "The United States condemns in the strongest possible terms todays attack . . . it was an outrageous violation of the ceasefire negotiated over the past several days, and of the assurances given to the United States and the United Nations. Hamas . . . must immediately and unconditionally release the missing Israeli soldier, and I call on those with influence over Hamas to reinforce this message . . . the international community must now redouble its efforts to end the tunnel and rocket attacks by Hamas terrorists on Israel and the suffering and loss of civilian life."[21]

White House Deputy National Security Advisor, Tony Blinken, further underscored the Obama Administration's condemnation of Hamas for violating the ceasefire: "This appears to be an absolutely outrageous action by Hamas, using the cover of a ceasefire to conduct a surprise attack through a tunnel, killing Israeli soldiers, and perhaps taking one hostage. We strongly, strongly condemn it. Israel has the right to defend itself, and it is obviously taking action to do so. But this is an outrageous action, and we look to the rest of the world to join us in condemning it, and those with influence on Hamas to use that influence to cease these actions."[22]

The blame for the new situation was placed unequivocally on Hamas. Continuing his condemnation of Hamas, President Obama said, "I think it's going to be very hard to put a ceasefire back together again if Israelis and the international community can't feel confident that Hamas can follow through on a ceasefire commitment." President Obama stated that the understandings included in the ceasefire gave Israel the right to continue its destruction of the tunnels. He affirmed that Israel was "entirely right" in insisting on this need, although it also advised that Israel should not advance its forces further toward populated areas.[23] The president further noted that the US was expressing its support for Israel not only in words, but also in deeds. For example, the US approved $225 million in emergency spending toward expansion of the Iron Dome system, after Prime Minister Netanyahu requested additional funding due to the ongoing conflict in Gaza. The US said it would help Israel guarantee its ability to defend its population as much as possible, and was maintaining close and ongoing contact with Israel at all levels.[24] The administration's stance was undoubtedly a key factor in the relatively broad legitimacy Israel received for its military actions throughout the operation. Prime Minister Netanyahu expressed this in one of his speeches: "We received international legitimacy from the global community . . . for very strong action against the terrorist organizations. This was substantial."[25]

While the president reiterated that the harm to civilians has to "weigh on our conscience," for the first time since the conflict began he made an effort to make it explicitly clear that he was well aware of the dilemmas facing Israel in its military operation. In his statement, President Obama continued: "On the one hand, Israel has a right to defend itself and it's got to be able to get at those rockets and those tunnel networks. On the other hand, because of the incredibly irresponsible actions on the part of Hamas to oftentimes house these rocket launchers right in the middle of civilian neighborhoods, we end up seeing people who had nothing to do with these rockets end up being hurt. Part of the reason why we've been pushing so hard for a ceasefire is precisely because it's hard to reconcile Israel's legitimate need to defend itself with our concern with those civilians."[26]

The president emphasized that the reality created did not deter the US from continuing its efforts to bring about a ceasefire, but that this task was now made more difficult than before by the fact that both Israel and the international community could no longer trust the promises of Hamas and its ability to control all the rival factions in the Gaza Strip.[27] It is quite possible that the change in tone by administration leaders was prompted by their great anger at Hamas for so crudely violating an express commitment given to the US administration in the context of the ceasefire. However, it would prove difficult to assess whether, and for how long, the administration would continue showing similar empathy for Israel's military actions in the Gaza Strip.

Indeed, this high level of support was short-lived. As related in Chapter 1, following the Al-Fakhura incident, where Israel responded to rocket fire from a UNRWA school in the Gazan Jabalia Camp and killed 40 Palestinians, the Obama Administration issued a sharp criticism of Israel's military activity. Without waiting for the result of the IDF inquiry into the strike, the State Department issued a condemnation, putting the blame for the harm to civilians on Israel.

According to the US Department of State statement, "The United States is appalled by today's disgraceful shelling outside an UNRWA school in Rafah sheltering some 3,000 displaced persons, in which at least ten more Palestinian civilians were tragically killed. The coordinates of the school, like all UN facilities in Gaza, have been repeatedly communicated to the Israeli Defense Forces. We once again stress that Israel must do more to meet its own standards and avoid civilian casualties. UN facilities, especially those sheltering civilians, must be protected, and must not be used as bases from which to launch attacks. The suspicion that militants are operating nearby does not justify strikes that put at risk the lives of so many innocent civilians . . . We continue to underscore that all parties must take all feasible precautions to prevent civilian casualties and protect the civilian population and comply with international humanitarian law."[28]

Although the Obama Administration was steadfast in its affirmation that Israel had the right to defend itself, its support was always qualified. By repeatedly emphasizing the need to exercise restraint and bring about a quick cessation of violence, the Obama Administration limited Israel's military freedom of action. This reflected a fundamental misunderstanding on the part of the Obama Administration, as well as many others following the ongoing situation from the West as to what was the actual root cause of the violence.

Keith Ellison, who was serving as the Democratic congressman from Minnesota's 5th district during Operation Protective Edge, penned an article in *The Washington Post* titled, "End the Gaza Blockade to Achieve Peace." Ellison, like Barack Obama, sympathized with the need for Israelis to live a life free from the danger of rocket fire, and acknowledged that Hamas must relinquish its arsenal for this to be possible, but simultaneously argued that Israel should make political concessions to Hamas to bring about peace, such as easing the blockade.[29] This argument contains numerous logical fallacies, noting that if the blockade were truly the root cause of violence between Israel and Hamas, then Egypt should be experiencing a similar level of violence, as it too has imposed a blockade on the costal enclave. What Ellison failed to understand is that Hamas is fighting Israel's very existence, not its blockade.

Former US President Jimmy Carter expressed a similar naivety when he co-authored an article for *Foreign Policy* with Mary Robinson. In the article, Carter and Robinson both urged the United States and the EU recognize Hamas as a legitimate political actor, stating that, "Only by recognizing its legitimacy as a political actor . . . can the West begin to provide the right incentives for Hamas to lay down its weapons."[30] The sentiments expressed by both Jimmy Carter and Keith Ellison exemplify how Hamas manipulates the media as a means of bolstering its support and legitimacy, in that Hamas can now claim that it has the support of Carter and Ellison, who both put the onus on Israel to appease Hamas. Recalling the charter on which Hamas was founded, a document saturated in anti-Semitic and genocidal rhetoric, stating that *jihad* is the only solution for the Palestinian people,[31] it is remarkably naïve to believe that peace between Israel and Hamas can be brokered merely by political concessions from Israel. When Israel made the ultimate political concession via its complete unilateral withdrawal from the Gaza Strip in 2005, it did not bring about peace, only rockets, terror tunnels, and war.

Mediation Efforts of Turkey and Qatar

Throughout the operation, administration officials had a tendency to distinguish between blame for the outbreak of the conflict – an issue that it underplayed – and blame for a lack of a ceasefire or agreement – an issue that it overplayed. There seemed to be a subtle message that even though Hamas was directly responsible for the outbreak of the conflict, Israel was not free of responsibility, since it had the opportunity to promote a settlement that would prevent conflict and failed to take advantage of it.

The administration also refrained from accepting Israel's request that serious discussion on policy issues would only take place *after* a ceasefire between Israel and Hamas had been established. During Kerry's visit to Cairo on July 21, 2014, he made it clear that a ceasefire, temporary or extended, would not last if fundamental problems were not addressed at the same time. The Secretary of State noted that the discussion on the substantive issues would begin "at some point," but he gave no details. Hamas had demanded the same thing.[32]

The following day, Kerry stated that, "Just reaching a ceasefire clearly is not enough. It is imperative that there be a serious engagement, discussion, negotiation regarding the underlying issues and addressing all of the concerns that have brought us to where we are today." When that would occur was not clear.[33] On another occasion, Kerry stated that the Palestinians can't have a ceasefire in which they think the status quo is going to stay and they're not going to have the ability to be able to begin to live and breathe more freely. In other words, the discussion on the substantive issues must take place during the fighting, just as Hamas demanded.[34]

In Secretary Kerry's autobiography, he initially remarks that Egypt appeared to be the most sensible country to broker negotiations between Israel and Hamas. However, at the same time he implicitly stated that Egypt was not really suitable to play the role of mediation in this conflict since both Israel and Egypt were working towards their "overriding desire to crush Hamas." Furthermore, "they were not even dealing with the elements of Hamas with the power to bring the war to a close." Secretary Kerry decided to seek out someone else who he believed, "had the leverage to force Hamas to stop firing rockets."[35]

On July 25, 2014, Kerry met in Paris with the Foreign Ministers of Turkey and Qatar, two countries that openly supported Hamas and its struggle against Israel. Qatar's financial support for Hamas is well known, as is Turkey's political support for the terror organization. John Kerry knew exactly who he was dealing with going into this meeting. The purpose of the meeting was to mobilize the two as key players in the efforts to achieve a ceasefire. It was clear that Israel would not be invited to the meeting. However, at the same time, the administration refrained from inviting Egypt or the Palestinian Authority, both of which have critical interests in any arrangement with Hamas. "Many Arab leaders," wrote Elliott Abrams, "were shocked to see Secretary of State Kerry in Paris with the Foreign Ministers of Qatar and Turkey, which were supporting Hamas, and without Egyptian or PA officials present." [36]

Specifically, Kerry's decision to meet with Turkey shocked Israel. Turkey and Israel had never maintained friendly relations, though there were periods of close cooperation between them primarily in security affairs. Following the Marmara incident, when Israel intercepted a Gaza bound flotilla of Turkish owned ships on May 31, 2010, relations between Turkey and Israel deteriorated. During the period that preceded the outbreak of hostilities, tensions were particularly high as Turkey loudly voiced its opposition to Israel's operation in Gaza. Campaigning for elections in August 2014, Turkish Prime Minister Recep Tayyip Erdoğan escalated his rhetoric against Israel to unite his base of support. He accused Israel of being more barbaric than Hitler[37] in its 'systematic genocide' of the Palestinians.[38] Turkey's government declared three days of mourning during Operation Protective Edge to show solidarity with the Palestinians in Gaza.[39] On July 18, Israel announced it was reducing diplomatic personnel in Turkey following violent protests in front of its embassy in Ankara.[40] Given the severe deterioration in bilateral relations between Turkey and Israel, Israel was dismayed at Kerry's invitation for Turkey to join the Paris meetings.[41]

After his meetings with the Turkish and Qatari Foreign Ministers, Kerry spoke in a firm if not threatening tone towards Israel. He noted the poor situation in which the Palestinian are living. There was no need to mention Israel as being responsible for their miserable conditions. It was absolutely clear to all: "I want everybody who cares about the Palestinian side to listen, and I want everybody in Israel to understand: we clearly understand – I understand that Palestinians need to live with dignity, with some – freedom . . . and they need a life that is free from the current restraints that they feel on a daily basis, and obviously free from violence."[42]

In order to show some balance in his position, Kerry again made the point for Israel's right to security: "Israelis," he said, "need to live free from rockets and from tunnels that threaten them, and every conversation we've had embraces a discussion about these competing interests that are a real for both. And so we need to have a solution that works at this . . . each side has powerful feelings about the history and why they are where they are. And what we're going to work at is how do we break through so that the needs are met and we have an ability to provide security for Israel and a future – economic and social and otherwise development for the Palestinians."[43]

These words further qualified the Obama Administration's previously unwavering support of Israel. There was no reference to Hamas' culpability for the outbreak of the conflict, to Israel's demand to demilitarize Gaza, or to Israel's right to monitor materials entering Gaza. Secretary Kerry presented the conflict as a clash of "competing interests that are real for both" the Palestinians and Israel. This implicitly placed Israel and Hamas on the same justification level as it suggested that the confrontation did not represent unjustified aggression by Hamas, as Israel claimed, but a struggle over "competing interests." Ultimately, from Israel's point of view, this conduct by the Obama Administration implied that it was seeking to push Israel into a corner and deny it the possibility of achieving the objectives for the operation that it had set for itself. Given this change in attitude, it was no surprise that the forthcoming settlement proposal reflected these positions and triggered a clash between the United States and Israel.[44]

On July 28, 2014, *Haaretz* obtained a US ceasefire draft for a weeklong ceasefire. According to *Haaretz* correspondent Barak Ravid, there were a number of elements that, from Israel's point of view, undermined its national interests:

1 There was almost no reference in the proposal to Israel's security needs, i.e., demilitarizing the Gaza Strip by removing rockets and heavy weapons and destroying the terror tunnels leading from Gaza to Israel. The emphasis was almost exclusively on Hamas' needs: opening the border crossings, allowing entry of goods and people, and transferring funds to Hamas to enable it to pay salaries.

2 According to the draft, the agreement was between the two parties, Israel and the "Palestinian factions," or in other words, Hamas and the other factions operating in the Gaza Strip. The two sides were of equal status.

3 The proposal did not give any status to the PA under Mahmoud Abbas. Not surprisingly, Israel's cabinet rejected the proposal. Wide circles in Israel, Egypt, and the United States harshly criticized the administration's conduct in the crisis, and in particular, the settlement proposal. It clearly did not reflect the "special relations" that supposedly exist between Israel and the United States.[45]

In Israel, cabinet ministers and senior officials were "in shock," with some referring to the proposal as a "prize for terror."[46] Ministers unanimously rejected the document. According to a *Haaretz* report, "Prime Minister Benjamin Netanyahu, Defense Minister Moshe Ya'alon, Justice Minister Tzipi Livni, and the rest of the security ministers could not believe what had been written down on paper."[47] The proposal was deemed so highly offensive that Israel decided not to issue an official announcement about the proposal "so as to avoid embarrassing the US Secretary of State and burning bridges at work." Instead, it was decided that Netanyahu would call Kerry personally and demand "significant improvements to the draft on matters essential to Israel."[48]

Israeli media echoed these concerns as well. Kerry was seen as a fool at best and as a sort of "disloyal traitor" at worst. David Horovitz, the founding editor of the *Times of Israel*, wrote, "What emerges from Kerry's self-initiated ceasefire mission . . . is that Jerusalem now regards him as duplicitous and dangerous."[49] He continued, "Whether through ineptitude, malice, or both, Kerry's intervention was not a case of America's top diplomat coming to our region to help ensure, through astute negotiation, the protection of a key ally. This was a betrayal."[50]

Barak Ravid, a prominent reporter for the left-leaning *Haaretz*, argued that Kerry's cease-fire proposal "might as well have been penned by [Hamas leader] Khaled Mashal."[51] Further elaborating, he argued that "The Secretary of State's draft empowered the most radical and problematic elements in the region – Qatar, Turkey, and Hamas – and was a slap on the face to the rapidly forming camp of Egypt, Israel, the

Palestinian Authority, Jordan, Saudi Arabia, and the United Arab Emirates, who have many shared interests. What Kerry's draft spells for the internal Palestinian political arena is even direr: it crowns Hamas and issues the Palestinian President Mahmoud Abbas with a death warrant."[52]

Indeed, the Palestinian Authority was also furious with this proposal. Speaking anonymously, a senior Palestinian official accused Kerry of trying "to destroy the Egyptian initiative and the Palestinian comments on it (the Abbas' plan) and present a substitute for our initiative . . . Kerry wanted to create a framework that would be an alternative to the Egyptian initiative and to our concept regarding it, in order to please Qatar and Turkey . . . Kerry wanted to take advantage of the war in order to again strengthen the Muslim Brotherhood's regional influence, because the Americans think – and will be proven wrong – that the moderate political Islam represented by the Muslim Brotherhood can combat radical Islam."[53]

In response, Obama Administration officials claimed that they had not expected the draft to be presented to the cabinet, and that Netanyahu's office had "breached protocol" by presenting it for a cabinet vote.[54] It is hard to believe that these claims were well received in Israel. There was no doubt that an important document such as this was carefully examined by the various government agencies and received the president's approval.

Obama Administration officials launched a major public relations campaign to "save face" and ease criticism. State Department spokeswoman Jen Psaki announced that the published proposal was not "a formal US proposal" but a "confidential draft."[55] It should be noted that John Kerry made the same claim in his autobiography.[56] In response to Israel's fierce criticism of Kerry's ceasefire draft, Psaki said that "It's simply not the way partners and allies treat each other," adding that, "those who want to support a ceasefire should focus on efforts to put it in place and not on efforts to criticize or attack one of the very people who are playing a prominent role in getting it done."[57]

Susan Rice, President Obama's National Security Advisor, also came to Kerry's defense: "I must tell you: we've been dismayed by some press reports in Israel mischaracterizing his efforts last week to achieve a ceasefire. The reality is that John Kerry on behalf of the United States has been working every step of the way with Israel. We'll continue to set the record straight when anyone distorts the fact."[58]

At the same time, and in order to display more sympathy toward Israel, the White House issued a memorandum on the main points of the conversation between President Obama and Prime Minister Netanyahu. The discussion included:

1 A "serious accusation" by the president against Hamas concerning its rocket fire and its use of tunnels to attack Israel;

2 Emphasis on the need to establish a humanitarian ceasefire, and then a permanent, *unconditional* ceasefire [emphasis added], as demanded by Israel;

3 Support by the United States for the Egyptian initiative, meaning that Turkey and Qatar were being excluded as key mediators, although administration spokesmen continued to emphasize the need to include the countries involved in the conflict and the regional actors in actions to reach a settlement;

4 An emphasis on the need to ensure Israel's security and strengthen the standing of the Palestinian Authority;

5 The concept that any permanent settlement of the conflict must ensure the disarming of the terrorist groups in Gaza and the demilitarization of Gaza. However, the president made clear that the issue of Gaza's demilitarization was not a matter for the immediate term, as Israel demanded, but something to be included in a comprehensive settlement of the Israeli–Palestinian conflict.[59]

National Security Advisor Susan Rice was also mobilized for the effort to improve the administration's image. At a meeting with Jewish leaders in the United States, she reiterated the administration's support for Israel: "As President Obama declared before the Israeli people in Jerusalem: 'So long as there is a United States of America . . . [Israel is] not alone.' That's why, from the moment that terrorist rockets began to rain down on Israel, this Administration, from President Obama on down, has made it clear: Israel has the same, unequivocal right to self-defense as every other nation. No nation can accept terrorists tunneling into its territory or rockets crashing down on its people." She also placed the blame clearly on Hamas, saying, "President Obama has been equally clear about who has been responsible for the violence. Hamas fired the rockets. Hamas deliberately targeted Israeli citizens, particularly civilians. Hamas refused an early plan for ceasefire. Hamas, in a time of glaring human need, instead of investing in the future of Gaza's children, built tunnels to kidnap and kill Israelis. So Hamas initiated this conflict. And, Hamas has dragged it on."[60]

The Secretary of State likewise repeated his deep commitment to Israel's security. Announcing his intention to pursue a ceasefire that completely reinforced Israel's right to self-defense, Kerry reaffirmed his commitment to Israel saying, "I've spent 29 years in the United States Senate and had a 100 percent voting record pro-Israel, and I will not take a second seat to anybody in my friendship or my devotion to the protection of the State of Israel."[61]

At a press conference on August 1, 2014, President Obama completed the campaign to defend Kerry. Rejecting the criticism of Secretary Kerry, Obama said, "Let me take this opportunity, by the way, to give Secretary John Kerry credit. He has been persistent. He has worked very hard. He has endured on many occasions really unfair criticism simply to try to get to the point where the killing stops and the underlying issues about Israel's security but also the concerns of Palestinians in Gaza can be addressed."[62]

US Punitive Measures Against Israel?

On July 21, 2014, the United States government suggested that Americans avoid travel to Israel if possible because of the increasingly deadly Israeli–Palestinian conflict in Gaza and the rockets fired into widespread parts of Israeli territory including cities. "The security environment remains complex in Israel, the West Bank, and Gaza, and US citizens need to be aware of the risks of travel to these areas because of the current conflict between Hamas and Israel," the State Department warning said. "Long-range rockets launched from Gaza since July 8, have reached many locations in Israel – including Tel Aviv, cities farther north, and throughout the south of the country," it stated. "Some rockets have reached Jerusalem and parts of the West Bank, including Bethlehem and Hebron."[63]

A day after, on July 22, two US airlines cancelled all flights to Israel until further notice, after a rocket landed near Tel Aviv's Ben-Gurion Airport. Delta Air Lines and United Airlines announced on July 23 that they were suspending service between the US and Israel indefinitely. US Airways scrapped its Tel Aviv service on the same day, and said that it was monitoring the situation in regard to future flights. Shortly after the airlines' announcement, the Federal Aviation Administration (FAA) told all US carriers they were prohibited from flying to Ben-Gurion Airport for 24 hours following a Hamas rocket explosion nearby. The rocket strike landed about one mile from the airport, the agency said. Delta Air Lines' one daily flight was already in the air when the company canceled the service. Delta said a Boeing 747 from New York was flying over the Mediterranean headed for Tel Aviv when it turned around and flew to Paris instead. Flight 468 had 273 passengers and 17 crewmembers on board.[64] Soon after the FAA decision, Lufthansa announced a 36-hour suspension that included subsidiaries German Wings, Austrian Airlines, and Swiss Airlines. Additionally, Air France announced its own indefinite suspension. The European Aviation Safety Agency issued a "strong recommendation" that airlines should avoid Tel Aviv. [65]

In response, Transportation Minister Israel Katz called on American aviation companies to return to normal functioning, stating that Ben-Gurion Airport was safe for takeoffs and landings. Furthermore, Minister Katz emphasized that there was no security concern for passenger planes: "There is no reason for the American companies to stop their flight and give a prize to terror," he said. Katz added that he believed the decision was an automatic reaction to the rocket landing, and hoped to convince them to reinstate flights on July 24. Israel noted the Iron Dome system had intercepted about 90 percent of the rockets launched from Gaza towards Israel. Hamas, on the other hand, stated that "The armed wing of the Hamas movement has decided to respond to the Israeli aggression, and we warn you against carrying out flights to Ben-Gurion Airport, which will be one of our targets today because it also hosts a military air base."[66]

According to a *Jerusalem Post* report, "tourism accounts for about 5 percent of Israel's exports and about 1.5–2 percent of GDP. Incoming tourism has already declined as a result of the rocket fire from Gaza, with organized groups canceling at a rate of about 30 percent between July and August. Yet the cancellation of flights, should it continue for a significant period of time, could have a greater impact on the economy. A May report by the Bank of Israel found that business travel to Israel tends to be more resilient than leisure tourism in the face of security problems. Without ways to get into the country, however, business travelers, who have historically accounted for 12–20 percent of travelers to Israel, will also be kept behind. Even worse, the precedent of flights canceled due to security may deter them from future business dealings."[67] However, beyond the damage to Israel's image, morale, and economy, Israel feared – justifiably – that closure of the airport would provide a very persuasive image of victory for Hamas.[68]

On July 23, 2014, the former Mayor of New York Michael Bloomberg came to Israel. He wished to demonstrate his criticism for the FAA decision to stop flights to Israel, and to show the American people that flying to Israel is safe. Prime Minister Netanyahu made a special gesture and came personally to the airport to welcome him. In his welcome speech, Netanyahu praised Mr. Bloomberg for his decision to come to Israel. "This is a clear proof that the flights to Israel are safe, we defend our airports. I think the decision of the FAA to stop flights to Israel was wrong and it just gave a prize to the Hamas and the terror. I call upon the FAA to resume flights to Israel immediately." Bloomberg praised the State of Israel as the only democracy in this part of the world. "Everyone can reflect his thinking openly. The Ben-Gurion Airport is the safest airport in the world. Israel has been under constant threat since its birth in 1948 and they know very well how to defend their state."[69]

On August 14 the *Wall Street Journal* reported that the administration was delaying a shipment of weapons to Israel after it found out that the weapons were being transferred solely on the basis of Pentagon approval.[70] Following harsh criticism of this unusual step – delaying weapons shipments to Israel during a military campaign – the State Department was quick to deny that this was a punitive measure against Israel. As proof, it referred to the fact that during the fighting the administration had transferred $225 million to Israel for the continued development of the Iron Dome system.[71]

According to the State Department, this was a routine bureaucratic move that is always taken when weapons are shipped to areas of tension, and did not reflect any change in policy toward Israel.[72] However, there was a widespread feeling in Israel that if the administration so desired, it had the tools to circumvent bureaucratic obstacles. This represented a departure from the longstanding position that the American government would not allow disagreements on the political level to harm the military-defense relationship with Israel. In any case, after intensive discussions with the United States, it was made clear that the supply of weapons would continue as usual.[73]

On the other hand, some Americans viewed the FAA's decision as an appropriate and justified security decision detached from politics. Given the fact that a missile struck down one mile from the runway, there was a very real, tangible, and direct threat to innocent civilians. According to aviation security consultant, Jeff Price, the FAA decision was a "prudent measure" as "the airline must protect their passengers and their asset (the airplane) from death, damage, and destruction, so they aren't going to fly into a location that they believe to be unsafe."[74]

Indeed, Richard Bloom, Director of Terrorism, Intelligence, and Security Studies at Embry-Riddle Aeronautical University, agreed: "Hamas has displayed some surprises – how many missiles they have and how far they can go. That explains why a number of (airlines) are getting out . . . it's extremely, extremely difficult to protect a commercial aircraft."[75] As Hamas posed a significant and real threat to airlines, the FAA acted responsibly in banning flights in order to prevent exposing American civilians to unnecessary danger.

Moreover, this decision was totally detached from politics. Before the FAA notice was issued, both Delta Air Lines and American Airlines made the independent decision to suspend service to Israel. The fact that private airlines initiated the ban on flights indicated that the decision was independent and informed by security needs, not political agendas. Furthermore, the FAA did not act unilaterally. Air Canada, Air France, Lufthansa, KLM, Brussels Airlines, Turkish Airlines, and the Russian carrier Aeroflot also suspended flights.[76] This suggests that the decision to suspend flights was not American political "retaliation," but a prudent security decision based on legitimate threats to travelers from around the world.

To Americans, Israel politicized the FAA decision. Immediately following the ban, the Israel Airport Authority urged airlines to reconsider, saying, "There is no reason that American carriers should stop flying to Israel and thus give a prize to terror."[77] However, political agendas and Israel's national interests do not and should not inform international security decisions.

Prime Minister Netanyahu also sprang into action, exerting political pressure to resume flights. He first thanked the UK Foreign Secretary, Philip Hammond, for British Airways continuing to fly to Ben-Gurion Airport: "I thank you for your moral focus and moral clarity and that's appreciated."[78] This statement conflated national security with international diplomacy and morality, laying the ground to wield political influence to overturn a security situation. Indeed, Prime Minister Netanyahu phoned Secretary of State Kerry seeking to reverse the FAA decision; however Kerry maintained that the US ban was solely due to safety concerns, adding that the decision would be reviewed within a day.[79]

While the ban was originally issued for security reasons, it may have been reversed for political reasons. After a day and a half, the US ended the ban on flights to Israel. State Department spokeswoman Marie Harf said, "There was significant new information about the threat – new information/intelligence about the threat which they took into account, which led to the rescinding of this notice, and also measures – new measures the government of Israel put in place to mitigate potential risks to civilian aircraft and aviation."[80] According to the State Department, the Israelis took concrete steps to improve certain safety protocols and procedures and shared sensitive information surrounding its defense capabilities. However, it's hard to imagine that Netanyahu's undisclosed remarks to Kerry did not influence the decision at all.

US Positions on Protective Edge – A Chronological Survey

The nature of the special relationship between the United States and Israel and how it was tested during Operation Protective Edge can be better understood within a chronological context, that is, by observing a survey of events and the American responses they generated throughout the conflict. There is perhaps no better place to start than with the event that was widely seen to have played a key role in igniting the 2014 Gaza War. On June 15, just three days after three Israeli teenagers were

kidnapped in the West Bank; Secretary of State John Kerry released the
following statement to the press:

> The United States strongly condemns the kidnapping of three Israeli
> teenagers and calls for their immediate release. Our thoughts and
> prayers are with their families. We hope for their quick and safe return
> home. We continue to offer our full support for Israel in its search for
> the missing teens, and we have encouraged full cooperation between the
> Israeli and Palestinian security services. We understand that coopera-
> tion is ongoing. We are still seeking details on the parties responsible
> for this despicable terrorist act, although many indications point to
> Hamas' involvement. As we gather this information, we reiterate our
> position that Hamas is a terrorist organization known for its attacks on
> innocent civilians and which has used kidnapping in the past.[81]

In response to the kidnapping, the Israel Defense Forces launched
Operation Brother's Keeper in search of the missing teens. However, on
June 30, the remains of Eyal Yifrach, Gilad Shaar, and Naftali Fraenkel
were discovered in a field northwest of Hebron. The murders sent shock-
waves throughout Israel, as tensions between Israelis and Palestinians
soared. Israel immediately placed blame on Hamas; however, this would
not be the only atrocity to take place before the war's onset, and the next
one would shift pressure onto Israel.

On July 2, just one day after the three murdered Israelis were buried,
Palestinian teenager Muhammad Hussein Abu Khdeir was kidnapped
and burned alive in an apparent revenge killing carried out by three
Israelis, further stoking the flames of conflict. The murder was strongly
condemned by the United States, as shown in the following press statement made by Secretary Kerry the same day:

> The United States condemns in the strongest possible terms the despi-
> cable and senseless abduction and murder of Muhammad Hussein Abu
> Khdeir. It is sickening to think of an innocent 17-year-old boy
> snatched off the streets and his life stolen from him and his family.
> There are no words to convey adequately our condolences to the
> Palestinian people.[82]

"The authorities", Kerry added, "are investigating this tragedy, a
number of Israeli and Palestinian officials have condemned it, and Prime
Minister Netanyahu has been emphatic in calling for all sides not to take
the law into their own hands." Kerry continued, "Those who undertake
acts of vengeance only destabilize an already explosive and emotional
situation. We look to both the Government of Israel and the Palestinian

Authority to take all necessary steps to prevent acts of violence and bring their perpetrators to justice. The world has too often learned the hard way that violence only leads to more violence and at this tense and dangerous moment, all parties must do everything in their power to protect the innocent and act with reasonableness and restraint, not recrimination and retribution."[83]

These events helped pave the way for the 2014 Gaza War, known in Israel as "Operation Protective Edge," which would officially begin on July 8. From the onset of the fighting, the United States voiced its support for Israel and condemnation of Hamas, but its support for Israel would always be qualified by concerns over civilian casualties. On July 8, Press Secretary Josh Earnest highlighted this concern during the daily press briefing:

> Let me start by saying that we strongly condemn the continuing rocket fire into Israel and the deliberate targeting of civilians by terrorist organizations in Gaza. No country can accept rocket fire aimed at civilians, and we support Israel's right to defend itself against these vicious attacks.[84]

He added, "At the same time, we appreciate the call that Prime Minister Netanyahu himself has made publicly to act responsibly. We're concerned about the safety and security of civilians on both sides. This means both the residents of southern Israel who are forced to live under rocket fire in their homes and the civilians in Gaza who are subjected to the conflict because of Hamas's violence."[85]

Additional press briefings held by Mr. Ernest on July 14 and July 15 continued on a similar note, urging restraint on all sides while expressing grave concerns over the rising Palestinian death toll.[86] Comments from Secretary of State John Kerry on July 15 echoed those made by Press Secretary Earnest, but further emphasized the importance of reaching a ceasefire:

> We urge all parties to support this ceasefire, and we support and we ask all the members of the Arab community, as they did yesterday at the Arab League meeting in Cairo, to continue to press to try to get Hamas to do the right thing here, which is cease the violence, engage in a legitimate negotiation, and protect the lives of people that they seem all too willing to put to risk.[87]

American President Barack Obama would express similar sentiments throughout the course of the war, defending Israel's right to self-defense while also emphasizing concern over rising Palestinian casualties and the need to achieve a ceasefire. On July 18, one day after Israel began its ground incursion into Gaza, President Barack Obama delivered the following statement:

> This morning, I spoke with Prime Minister Netanyahu of Israel about the situation in Gaza. We discussed Israel's military operation in Gaza, including its efforts to stop the threat of terrorist infiltration through tunnels into Israel. I reaffirmed my strong support for Israel's right to defend itself. No nation should accept rockets being fired into its borders, or terrorists tunneling into its territory. In fact, while I was having the conversation with Prime Minister Netanyahu, sirens went off in Tel Aviv.[88]

The president added that the United States and its friends and allies "are deeply concerned about the risks of further escalation and the loss of more innocent life. And that's why we've indicated, although we support military efforts by the Israelis to make sure that rockets are not being fired into their territory, we also have said that our understanding is the current military ground operations are designed to deal with the tunnels, and we are hopeful that Israel will continue to approach this process in a way that minimizes civilian casualties and that all of us are working hard to return to the ceasefire that was reached in November of 2012."[89]

An almost identical statement made by President Obama on July 21, on the South Lawn of the White House, as President Obama urged the international community to concentrate its efforts on bringing about an end to the fighting.[90] The statement followed a phone call with Israel's Prime Minister Benjamin Netanyahu that occurred on July 20, in which the president discussed Israel's ongoing military operation, reiterated the United States' condemnation of attacks by Hamas against Israel, and reaffirmed Israel's right to defend itself, while also raising serious concerns about the growing number of casualties, including increasing Palestinian civilian deaths in Gaza and the loss of Israeli soldiers.[91]

Meanwhile, Secretary of Kerry offered his perspective during a series of interviews on July 20. One of these interviews was with ABC News' George Stephanopoulos, in which Secretary Kerry appeared to pin most of the blame for the violence on Hamas: "What they need to do," he stated, "is stop rocketing Israel and accept a ceasefire. It's very, very clear that they've tunneled under Israel. They've tried to come out of those tunnels with people with handcuffs and tranquilizer drugs to capture Israeli citizens and hold them for ransom, or worse. They've been rocketing Israel with thousands of rockets. They've been offered a ceasefire, and they've refused to take the ceasefire. Even though Egypt and others have called for that ceasefire, they've just stubbornly invited further efforts to try to defuse the ability to be able to rocket Israel. So it's ugly, obviously. War is ugly, and bad things are going to happen. But they need to recognize their own responsibility. We have offered to have a ceasefire and then negotiate the issues."[92]

Later that day, an interview with Chris Wallace from Fox News Sunday brought up an embarrassing open microphone "gaffe" on the part of John Kerry, as he was caught in-between interviews speaking with a top aide over the phone. The two were discussing the situation in Israel and the fact that 14 Israelis had either been shot or killed during an operation, completely unaware of the fact that the conversation was being recorded. "It's a hell of a pinpoint operation," Kerry was caught saying over the phone.[93] Chris Wallace replayed the clip for Secretary Kerry during their interview, and then pressed him as to whether or not he believed that the Israelis were going too far.

"It's tough to have this kind of operation," Kerry said. "And I reacted obviously in a way that anybody does with respect to young children and civilians. But war is tough, and I said that publicly and I'll say it again. We defend Israel's right to do what it is doing in order to get at those tunnels. Israel has accepted a unilateral ceasefire. It's accepted the Egyptian plan, which we also support. And it is important for Hamas to now step up and be reasonable and understand that you accept a ceasefire, you save lives, and that's the way we can proceed to have a discussion about all of the underlying issues which President Obama has clearly indicated a willingness to do."[94]

Secretary Kerry's frustration was clearly visible. Another interview that day conducted by Candy Crowley from CNN also inquired as to whether or not Kerry felt that Israeli actions in Gaza were going too far. Secretary Kerry was careful in his responses: "No country," he added, "no human being is comfortable with children being killed, with people being killed, but we're not comfortable with Israeli soldiers being killed either, or with people being rocketed in Israel. So in war, it's very difficult. There tends not to be a sort of equilibrium in terms of these things. The fact is that we've asked Israel and Israel has said we will try to reduce whatever we can with respect to civilian involvement, and civilians have been warned to move well ahead of time. The fact is that Hamas uses civilians as shields and they fire from a home and draw the fire into the home, precisely to elicit the kind of question you just asked. We need to have a ceasefire."[95]

At this point in the conflict, Western media outlets were already fixated on Palestinian death tolls, and this would be a common theme in White House press briefings throughout the duration of Operation Protective Edge. During the press briefing on July 21, Press Secretary

Earnest was repeatedly pressed on whether or not the United States considered Israel to be acting with disproportionate force. In response, Earnest reiterated Israel's right to self-defense while explaining how the tactics used by Hamas are specifically meant to increase civilian casualties, but also admitted that the United States desired to see Israel take "greater steps" to ensure the safety of civilians.[96]

By July 21, Secretary Kerry had arrived in the Middle East, making stops in Cairo and Ramallah over a period of several days. Remarks given by Secretary Kerry in Cairo praised the United States' decision to provide $47 million in direct humanitarian assistance to the Gaza Strip, while also affirming his support for the Egyptian initiative for the ceasefire.[97] However, future developments regarding Kerry's diplomatic actions would bring many to doubt this support, particularly in Israel, Egypt, and the Palestinian Authority. Nevertheless, on July 23, 2014, Secretary Kerry met with Palestinian President Mahmoud Abbas in Ramallah, and delivered the following remarks:

> I have been in constant touch with President Abbas and the Palestinian Authority over the course of the last months. But particularly in the last days, we have been talking about how to achieve an end to the current violence and an effort to try to not only have a ceasefire, but build a process that can create a sustainable way forward for everybody. I'm very grateful to President Abbas for his leadership, for his deep engagement in the effort to try to find a ceasefire. He has traveled tirelessly, he has been working with all of the interested groups and parties, and encouraging people to do the responsible thing, which is to come to the table – not only have a ceasefire, but then negotiate the immediate issues and the underlying issues.[98]

"We had a good conversation today about how we can take further steps," Kerry added, "and we're doing this for one simple reason: The people in the Palestinian territories, the people in Israel, are all living under the threat or reality of immediate violence, and this needs to end for everybody. We need to find a way forward that works, and it's not violence. President Abbas has been committed to nonviolence and committed to a harder route. Sometimes it's very satisfying for people to see the immediate impact of the violence, but it doesn't take you to a solution. President Abbas understands the road to the solution, and that's what we're working for. So we will continue to push for this ceasefire. We will continue to work with President Abbas and others in the region in order to achieve it. And I can tell you that we have, in the last 24 hours, made some progress in moving towards that goal. And I will leave here now with President Abbas' thoughts about how we could make some progress, and I will go and meet with Prime Minister Netanyahu and subsequently return to Cairo, where we will continue in the hopes that before long, we can change course and, for everybody's sake, end this violence and move to a sustainable program for the future."[99]

After his meeting in Ramallah, Secretary Kerry travelled back to Cairo where he delivered his final remarks on July 25, just before leaving the region for Paris, where he was scheduled to meet with his Qatari and Turkish counterparts. In opening, Secretary Kerry expressed his gratitude for the Egyptians, UN Secretary-General Ban Ki-moon, and Palestinian Authority President Mahmoud Abbas for their efforts and their desires to bring about peace. Kerry continued:

> Let me just say that the agony of the events on the ground in Gaza, the West Bank, and Israel, all of them together, simply cannot be overstated. The daily reality for too many people of grief and blood and loss and tears, it all joins together to pull at the fabric of daily life in each of their communities. In Israel, millions of people are living under constant threat of Hamas rocket fire and tunnel attacks, and they're ready to take cover at any moment's notice. And I've had telephone conversations with the Prime Minister interrupted by that fact. Earlier this week I had a chance to visit with the family of a young man by the name of Max Steinberg, an American – one of two Americans killed in this devastating conflict – and the mother of Naftali Fraenkel, who was murdered at the outset – whose son was murdered at the very outset of this crisis.[100]

"So any parent in the world, regardless of somebody's background," Kerry added, "can understand the horror of losing a child or of seeing these children who are caught in the crossfire. In Gaza, hundreds of Palestinians have died over the past few weeks, including a tragic number of civilians. And we've all read the headlines and seen the images of the devastation: 16 people killed and more than 200 injured in just a single attack yesterday; women and children being wheeled away on stretchers; medics pulling shrapnel out of an infant's back; a father nursing his three-year-old son. The whole world is watching tragic moment after tragic moment unfold and wondering: When is everybody going to come to their senses?"[101]

"Both the Israelis and the Palestinians," Kerry concluded, "deserve and need to lead normal lives, and it's time for everyone to recognize that violence breeds violence and that the short-term tactical gains that may be made through a violent means simply will not inspire the long-term change that is necessary and that both parties really want."[102]

These comments are significant because they implicitly placed Israel and Hamas on the same playing field. It had been obvious that Israel did not want this conflict, as shown by the eleven ceasefires that Israel accepted.[103] On the other hand, Hamas clearly showed its desire for conflict in their refusals to cease hostilities and their outright rejection or violation of each and every ceasefire proposal until the conclusion of the fighting on August 26, 2014. Furthermore, it is estimated that had Hamas accepted the ceasefire proposed on July 15, just prior to the Israeli ground invasion, that approximately 90 percent of Gazan casualties could have been avoided.[104]

Given this information, it would be highly illogical to place Hamas and Israel on the same playing field when so much of the suffering in Gaza Strip can be directly attributed to the irresponsible behavior and decision making of those running the Palestinian enclave. The United States knew (or at least should have known) this, and that there was absolutely no moral equivalence between the two sides. Yet while the Obama Administration occasionally mentioned or alluded to the culpability of Hamas, this was a point that was never fully driven home. Instead, the administration repeatedly subverted this point by continuing to imply that Israel was also partially to blame for the fighting.

The remainder of Secretary Kerry's comments that day would focus primarily on the efforts being made to achieve a ceasefire, but it was clear at this point that distrust between Israel and the Obama Administration was starting to culminate. From Cairo, John Kerry travelled to Paris to meet the Foreign Ministers of Turkey and Qatar. He delivered the following statement on July 26:

> I want anybody who cares about the Palestinian side of these issues to listen, and I want everybody in Israel to understand: we clearly understand, I understand, that Palestinians need to live with dignity, with some freedom, with goods that can come in and out, and they need a life that is free from the current restraints that they feel on a daily basis, and obviously free from violence. But at the same time, Israelis need to live free from rockets and from tunnels that threaten them, and every conversation we've had embraces a discussion about these competing interests that are real for both. And so we need to have a solution that works at this.[105]

"I understand," he added, "that Israel can't have a ceasefire in which they are not able to – [eliminate the tunnel threat]. The tunnels have to be dealt with. We understand that; we're working at that. By the same token, the Palestinians can't have a ceasefire in which they think the status quo is going to stay and they're not going to have the ability to be able to begin to live and breathe more freely and move within the crossings and begin to have goods and services that come in from outside. These are important considerations. Each side has powerful feelings about the history and why they are where they are. And what we're going to work at is how do we break through that so that the needs are met and we have an ability to provide security for Israel and a future – economic and social and otherwise – development for the Palestinians. That's what this is about."[106]

The statement notably omits any mention of Hamas, or its culpability for the conflict, implicitly placing the actions of Israel and the terrorist organization Hamas on a level playing field. Additionally, this meeting along with the draft ceasefire it produced would soon become a significant source of controversy for the Obama Administration. Widespread condemnation of Secretary Kerry's actions would emanate out of both Israel and the Palestinian Authority, as Turkey and Qatar were both outspoken supporters of Hamas. Again, the fact that both Israel and the Palestinian Authority condemned Secretary Kerry's actions should not be understated, as very rarely do these entities find themselves in such agreement. In Israel, the draft-ceasefire was being called a "prize for terror,"[107] and officials from the Palestinian Authority were accusing the United States of trying to destroy the Egyptian initiative and Palestinian Authority comments on it as a means of appeasing Turkey and Qatar.[108]

Meanwhile, on July 27, President Obama and Prime Minister Netanyahu spoke again over the phone to discuss the situation in Gaza. Again, the president reaffirmed Israel's "right to defend itself," while reiterating the United States' "serious and growing concern" over the rising number of Palestinian civilian deaths and the loss of Israeli lives, in addition to the worsening humanitarian situation in Gaza. Additional points of emphasis included the strategic imperative of instituting an immediate, unconditional humanitarian ceasefire, a return to the 2012 ceasefire agreement between Israel and Hamas, the eventual disarmament and demilitarization of the Gaza Strip, and support for the Egyptian initiative to bring these goals to fruition.[109]

Despite its voiced support for the Egyptian initiative, it seemed that American diplomatic actions were actively subverting the initiative. Secretary Kerry was already coming under intense scrutiny in Israel for both his meeting with the Qataris and Turks in Paris and the leaked ceasefire draft, a topic that was featured in the White House's daily press briefing on July 29. Press Secretary Earnest defended Secretary Kerry's credentials, stating:

Secretary Kerry, as Mr. Blinken noted yesterday, is a strong defender of our allies in Israel. And that is why I guess I would be so bold as to suggest that it is in the interest of the Israeli people for the harsh words that we've seen directed at the Secretary not affect his ability to continue to be a strong advocate for them. Secretary Kerry has worked doggedly over the last year or so since he took office – I guess it has been a little longer than that – pressing both sides in terms of a broader – to the negotiating table in search of a broader peace agreement. What he has been engaged in more recently is working with Palestinian leaders, Israeli leaders, other leaders in the Arab world, UN officials, to try to put in place a permanent cessation of hostilities based on the November 2012 ceasefire agreement.[110]

Following up with regard to the leaked draft proposal and the backlash it generated in Israel, Earnest remarked:

The facts are that the cease-fire proposal that was put forward by the Egyptians two weeks ago – this is the ceasefire agreement that Israel readily accepted – included many of the elements that some anonymous Israeli officials are now suggesting were wrongly included in the document that was circulated by Secretary Kerry. That's the second thing that's important to understand, is this is a document that was circulated among the parties that reflected an attempt to get a dialogue going between the parties. This did not reflect a specific American proposal. This reflected an effort to try to find some common ground and to elicit comment from the Israelis to try to find the kind of ceasefire agreement that they would believe would be in their best interests, and would also provide for greater security of their citizens.[111]

"So," he added, "this is part of the diplomatic effort that was underway, and it is in line with the proposal that the Israelis had readily agreed to a couple of weeks ago. So those facts as it relates to the document that we circulated by Secretary Kerry are really important in this case. In terms of the broader relationship, again, we've said for a couple of days now that those comments were pretty disappointing. But our determination, and more importantly, Secretary Kerry's determination to try to put in place an immediate ceasefire that would end the crossfire that so many civilians – innocent civilians are caught in the middle of right now continues to be a top priority. And I know that he's working very hard as we speak in pursuit of that agreement."[112]

A few days later, as the result of the shelling of a UN school in Gaza, the pressure was shifted back onto Israel. On July 31, Press Secretary Earnest challenged Israel to do more to protect civilian lives in Gaza. During a press briefing, Earnest stated that, "the shelling of a UN facility that is housing innocent civilians who are fleeing violence is totally unacceptable and totally indefensible. And it is clear that we need our allies in

Israel to do more to live up to the high standards that they have set for themselves."[113] Brief mention was given to the fact that the Israeli government acknowledged firing in the area in response to fire from Hamas in the immediate vicinity of the school, but emphasis was not given to the fact these facilities have repeatedly been used to shield Hamas' weapons and personnel.

On July 31 the Office of the Press Secretary announced a humanitarian ceasefire set to begin at 8:00 a.m. on August 1, and emphasized the United States' support, urging all parties to act with restraint.[114] The ceasefire would be short-lived. An attack on August 1 that led to the killing of two Israeli soldiers and the apparent abduction of another was condemned in a statement made by Secretary Kerry the same day, adding that, "It was an outrageous violation of the ceasefire negotiated over the past several days, and of the assurances given to the United States and the United Nations."[115]

On August 4 the shelling of a UN facility in Gaza was again a significant subject of conversation during the White House press briefing, with Secretary Earnest relaying the State Department message that "the suspicion that militants are operating nearby does not justify strikes that put at risk the lives of so many innocent civilians," but also emphasized that "there have been reports of Hamas using innocent civilians as cover to protect their weapon stockpiles or even to protect Hamas fighters. That is a deplorable technique or strategy, and it is one that we have strongly condemned and continue to condemn."[116] Press Secretary Earnest also took this time to emphasize the strength of the relationship between Israel and the United States despite turbulence surrounding the conflict in Gaza.

During an interview with BBC on August 5, John Kerry was pressed as to whether Washington fully supported Israel's offensive in Gaza, to which Kerry replied, "We fully support Israel's right to defend itself and the fact that it was under attack by rockets, by tunnels, and it had to take action against Hamas. Hamas has behaved in the most unbelievably shocking manner of engaging in this activity. And yes, there has been horrible collateral damage as a result of that, which is why the United States worked very, very hard with our partners in the region, with Israel, with Egyptians, with the Palestinian Authority, with President Abbas, to try to move towards a ceasefire."[117]

On August 6 President Obama was speaking at a press conference following the US–Africa Leaders Summit when he was asked whether or not he agreed that Israel's operation in Gaza was "justified and proportionate," and what role the US would play in the peace talks: "I have said from the beginning," President Obama remarked, "that no country would tolerate rockets being launched into their cities. And as a consequence, I have consistently supported Israel's right to defend itself, and that includes doing what it needs to do to prevent rockets from landing on population centers and, more recently, as we learned, preventing tunnels from being dug under their territory that can be used to launch terrorist attacks. I also think it is important to remember that Hamas acts extraordinarily irresponsibly when it is deliberately siting rocket launchers in population centers, putting populations at risk because of that particular military strategy."[118]

"Now, having said all that," President Obama continued, "I've also expressed my distress at what's happened to innocent civilians, including women and children, during the course of this process. And I'm very glad that we have at least temporarily achieved a ceasefire. The question is now how do we build on this temporary cessation of violence and move forward in a sustainable way. We intend to support the process that's taking place in Egypt." President Obama continued, "I think the short-term goal has to be to make sure that rocket launches do not resume, that the work that the Israeli government did in closing off these tunnels has been completed, and that we are now in the process of helping to rebuild a Gaza that's been really badly damaged as a consequence of this conflict. Long term, there has to be a recognition that Gaza cannot sustain itself permanently closed off from the world and incapable of providing some opportunity – jobs, economic growth – for the population that lives there, particularly given how dense that population is, how young that population is."[119]

"We're going to have to see a shift in opportunity for the people of Gaza," President Obama added. "I have no sympathy for Hamas. I have great sympathy for ordinary people who are struggling within Gaza. And the question then becomes, can we find a formula in which Israel has greater assurance that Gaza will not be a launching pad for further attacks, perhaps more dangerous attacks as technology develops into their country. But at the same time, ordinary Palestinians have some prospects for an opening of Gaza so that they do not feel walled off and incapable of pursuing basic prosperity."[120]

"I think," the president continued, "there are formulas that are available, but they're going to require risks on the part of political leaders. They're going to require a slow rebuilding of trust, which is obviously very difficult in the aftermath of the kind of violence that we've seen. So I don't think we get there right away, but the US goal right now would be to make sure that the ceasefire holds, that Gaza can begin the process of rebuilding, and that some measures are taken so that the people of Gaza feel some sense of hope, and the people of Israel feel confident that they're not going to have a repeat of the kind of rocket launches that we've seen over the last several weeks."[121]

"And Secretary Kerry," President Obama said, "has been in consistent contact with all the parties involved. We expect we will continue to be trying to work as diligently as we can to move the process forward. It is also going to need to involve the Palestinian leadership in the West Bank. I have no sympathy for Hamas. I have great sympathy for some of the work that has been done in cooperation with Israel and the international community by the Palestinian Authority. And they've shown themselves to be responsible. They have recognized Israel. They are prepared to move forward to arrive at a two-state solution. I think Abu Mazen is sincere in his desire for peace. But they have also been weakened, I think, during this process. The populations in the West Bank may have also lost confidence or lost a sense of hope in terms of how to move forward. We have to rebuild that, as well. And they are the delegation that's leading the Palestinian negotiators. And my hope is, is that we'll be engaging with them to try to move what has been a very tragic situation over the last several weeks into a more constructive path."[122]

President Obama would speak with King Abdullah II of Jordan over the phone on August 8, as they discussed the need for an immediate cessation of hostilities and a durable ceasefire in Gaza, as well as increased support to civilians in Gaza who have suffered tremendously during the conflict.[123] Despite enormous efforts to establish a ceasefire, the fighting would continue until August 26, when a ceasefire was finally agreed upon, thus concluding Operation Protective Edge. The same day that the ceasefire was announced, Secretary of State John Kerry announced the United States' strong support for the ceasefire agreement.[124]

Almost a full month after Operation Protective Edge came to an end, President Obama spoke to the United Nations General Assembly on September 24:

> The status quo in the West Bank and Gaza is not sustainable. We cannot afford to turn away from this effort – not when rockets are fired at innocent Israelis, or the lives of so many Palestinian children are taken from us in Gaza. So long as I am President, we will stand up for the principle that Israelis, Palestinians, the region and the world will be more just and more safe with two states living side by side, in peace and security. So this is what America is prepared to do: Taking action against immediate threats, while pursuing a world in which the need for such action is diminished. The United States will never shy away from defending our interests, but we will also not shy away from the promise of this institution and its Universal Declaration of Human Rights – the notion that peace is not merely the absence of war, but the presence of a better life.[125]

"I realize," President Obama added, "that America's critics will be quick to point out that at times we too have failed to live up to our ideals; that America has plenty of problems within its own borders. This is true. In a summer marked by instability in the Middle East and Eastern Europe, I know the world also took notice of the small American city of Ferguson, Missouri – where a young man was killed, and a community was divided. So, yes, we have our own racial and ethnic tensions. And like every country, we continually wrestle with how to reconcile the vast changes wrought by globalization and greater diversity with the traditions that we hold dear."[126]

Shortly after, on October 1, President Obama met with Israeli Prime Minister Benjamin Netanyahu in the Oval Office, where the two delivered remarks prior to their bilateral meeting. "We meet at a challenging time," Obama remarked, "Israel is obviously in a very turbulent neighborhood, and this gives us an opportunity once again to reaffirm the unbreakable bond between the United States and Israel, and our ironclad commitment to making sure that Israel is secure." President Obama continued:

> Throughout the summer, obviously all of us were deeply concerned about the situation in Gaza. I think the American people should be very proud of the contributions that we made to the Iron Dome program to protect the lives of Israelis at a time when rockets were pouring into Israel on a regular basis. I think we also recognize that we have to find ways to change the status quo so that both Israeli citizens are safe in their own homes and schoolchildren in their schools from the possibility of rocket fire, but also that we don't have the tragedy of Palestinian children being killed as well.[127]

"And so," President Obama continued, "we'll discuss extensively both the situation of rebuilding Gaza but also how can we find a more sustainable peace between Israelis and Palestinians. Our agenda will be broader than that, obviously. I'll debrief Bibi on the work that we're doing to degrade and ultimately destroy ISIS, and the broader agenda that I discussed at the United Nations, which is mobilizing a coalition not only for military action, but also to bring about a shift in Arab states and Muslim countries that isolate the cancer of violent extremism that is so pernicious and ultimately has killed more Muslims than anything else."[128]

Throughout Operation Protective Edge, the United States maintained close contacts and cooperation with Israel despite moments of significant turbulence, testifying to the strength of the special relationship between Israel and the United States. That being said, the United

States support for Israel's military action in Gaza was qualified from the beginning, and therefore limited the IDF's freedom of action. As a result, the IDF was never even close to utilizing its full capabilities, which frustrated Israeli efforts to achieve their objective. Nevertheless, after a significant amount of time and resources, the tunnel threat was destroyed and deterrence was reinstated, but so long as Hamas remains a significant and hostile threat on Israel's southern border, the threat of renewed conflict remains.

Conclusion

Operation Protective Edge began and took place under political conditions that seemed to be favorable, if not ideal for Israel. However, while Israel enjoyed overwhelming military superiority over Hamas, it lacked the legitimacy to fully utilize its military power. Nevertheless, numerous members of the international community, including Canada, the US, the EU, and Australia, all acknowledged Israel's right to use force in the context of the 2014 Gaza War,[1] after all, Israel is an advanced, pro-Western democracy and unofficial US ally (major non-NATO ally) defending itself against a terrorist organization that had been outlawed by Congressional legislation for operating in contravention of international law.[2]

Hamas initiated the conflict by its involvement in the kidnapping and murder of three Israeli teenagers. Moreover, Israel, unlike Hamas, demonstrated a willingness to bring about an end to hostilities by accepting every ceasefire proposal made throughout the operation. Israel also kept strictly to the agreements and understandings previously reached with Hamas. While Israel undertook extensive measures to avoid killing or wounding civilians, Hamas was engaged in deliberate and indiscriminate firing towards civilian targets in Israel, an action that according to the Obama Administration spokespersons was "completely unacceptable." [3]

Additionally, the public and brutal executions Hamas conducted during the operation against men who were allegedly supporting Israel damaged its image as an organization that focuses on social welfare and enjoys widespread public sympathy. Its comparison to ISIS in global public opinion was inevitable, even though the United States has not yet officially accepted Prime Minister Netanyahu's comparison between the two organizations.

Under these circumstances it could have been expected that during the conflict the Obama Administration would have fully backed Israel, but in reality the situation was entirely different. When administration officials referred to the conflict, they often projected the message that Israel and Hamas were two sides fighting each other as equals, and that the administration was not favoring either of them. The overriding goal was to end the conflict, or in other words, bring about calm on the basis of the understandings that led to the end of Operation Pillar of Defense in 2012. Secretary Kerry expressed this poignantly when, toward the end of Operation Protective Edge, he was asked directly whether the United States gave its full support to Israel in the operation. He refrained from answering in the affirmative, making do with a routine statement to the effect that the United States supported Israel's right to defend itself.[4]

The administration's somewhat alienated stance toward Israel during the operation was likely dictated by certain strategic considerations. The administration acknowledged that it was Hamas that initiated the latest conflict in Gaza, and was well aware of the fact that the Israeli government, unlike Hamas, demonstrated a sincere desire throughout the campaign to agree to a ceasefire and to return to a state of calm. Nevertheless, in comments made by administration officials, there was a tendency to make clear, albeit implicitly, that Israel was also responsible for the outbreak of the conflict.[5]

As is well known, the United States has been opposed, in various levels of intensity, to Israel's settlement policy in the West Bank. It should be recalled that President Obama was determined to bring an end to this policy, which he regarded as a major obstacle to the conclusion of a peace agreement between Israel and the Palestinians. Shortly after coming to office, President Obama made it clear that he was not going to follow his predecessors with regard to Israel's settlement policy: "Part of being a good friend," he said, "is being honest. And I think there have been times where we are not as honest as we should be about the fact that the current direction, the current trajectory in the region, is profoundly negative–not only for Israeli interests but also US interests. And that's part of a new dialogue that I'd like to see encouraged in the region."[6]

Shortly afterwards, during his Cairo speech on June 5, 2009, President Obama made clear his view that he would not tolerate any longer the existing Israeli settlement policy: "The United States," he said, "did not accept the legitimacy of the settlements." "Construction in the settlements," the president stated, "was a violation of previous agreements and undermined efforts to achieve peace. Settlements have to stop."[7] The leadership in Israel liked neither the demands of the president nor the way they were presented. While Obama's tone toward the Palestinians was moderate and cautious, he was much more forceful and resolute toward Israel. Palestinians were given vague guidelines and subjective demands, while Israelis were asked to prove they had made concrete, practical steps. Prime Minister Netanyahu rejected President Obama's demands to completely freeze the settlement policy. Eventually, Israel and the United States agreed on a ten-month freeze of the settlement build-up in the West Bank. This did not satisfy

President Obama, and he often blamed Israel and its settlement policy for the lack of advancement in the peace process, and implicitly – for the outbreak of hostilities.

Furthermore, the US had to take into consideration the intensive Turkish and Qatari involvement in the conflict. These two economically and politically powerful countries, which have a close relationship with the United States, have openly declared their support for Hamas. This fact greatly limited the US administration's ability to maneuver during the campaign. It likely estimated that if it were to express explicit support for Israel and Egypt, this could engender a harsh response from Turkey and Qatar that would harm essential US interests. The turbulence in today's Middle East and the need to deploy the US military in operational tasks in the Middle East, made it necessary for the administration to avoid a crisis with these two important countries. This is presumably the reason why the administration attempted to make Turkey and Qatar key actors in the mediation efforts, knowing well that those two states adopt hostile positions towards Israel. Naturally, the American attempt to give Turkey and Qatar a dominant role in the negotiations brought about criticism on the part of Israel. Only after it had been harshly criticized did the administration renege on the move.

Another consideration that contributed to the shaping of US policy during the conflict was the exclusion of the United States from the process that was intended to bring an end to the warfare through a set of understandings between Israel and Hamas at the initial stages of the conflict. During the campaign, the United States found itself in the embarrassing position of lacking a meaningful position or policy in the process of achieving a ceasefire and regulating relations between Egypt, Israel, and Hamas. Most importantly, this was the first war since the establishment of the State of Israel in which the United States did not play a dominant role in the process of achieving a settlement of hostilities. Its attempts to be part of the efforts at a settlement involved incidents that were both embarrassing to itself and to its representatives. Ultimately, the administration had no choice but to accept the fact that Egypt was leading the process of reaching an arrangement with Hamas. Among various circles in the administration, the prevailing assumption was that Netanyahu, estimating that American engagement in the diplomatic process to end the war, would never serve Israel's best interests, had pushed the United States aside. The unfriendly policy of the United States towards Israel has been interpreted as retaliation against this Israeli policy.[8]

Finally, the US had to face the issue of image and its impact on public opinion during the conflict: The harrowing photographs from Gaza publicized by the global media aggravated Israel's image problem. For the administration, it was especially difficult to accept the sight of injured children and harm to civilians within, or next to UN institutions. The administration was familiar with Israel's explanations and even voiced them a number of times, but the pictures made it difficult for the administration to express wide support for Israel.

The underlying conclusion is that throughout the operation, the United States did not treat Israel as one would expect for a close ally. From Israel's point of view, it seemed clear that the US was not acting in a way that would allow Israel to defeat Hamas. On the contrary, the Obama Administration seemed determined to avoid the defeat of Hamas. This point of view suits the long-standing conception in Israel with regard to relations with the United States, which was established during the 1950s, mainly by Israel's first Prime Minister, David Ben-Gurion. The bottom line of this thinking is that Israel views the United States as major strategic asset, and will do its utmost to strengthen relations with the United States. However, Israel will never be ready to see the relationship with the United States as an alternative to the build-up of its own strategic power. The United States, Ben-Gurion noted, is a world leading power, and it has a variety of international interests. "We can never know if it will fulfill its commitments to Israel when this comes to the test." The relationship with the United States is an addition, not a substitute, to Israel's self-power.

However, and with a broad perspective, it is important to set out the following conclusion. Although throughout Operation Protective Edge the Obama Administration did not treat Israel as a close ally, and despite the pitfalls, disagreements, arguments, and mutual insults between Israel and the United States during the operation, the overall picture that emerges is that the "special relationship" remained stable and successfully survived the severe turmoil surrounding Operation Protective Edge. Throughout the operation, Israel and the United States conducted an ongoing, intensive, deep, and intimate dialogue, as befits countries with a broad strategic partnership. Furthermore, throughout the operation, there was an effort by both sides to avoid a rupture, with a clear emphasis on continuing an intensive dialogue in spite of the disagreements.

Moreover, it is impossible to ignore the fact that even when the administration chose to publicly or discreetly emphasize its displeasure with Israel's conduct during the campaign, it avoided placing heavy pressure on Israel to change the nature of its military operation. This means that in practice, throughout the operation, i.e., a period just short of two months, the United States allowed Israel a fairly large freedom of action even when Israel's military actions were unprecedented and very far from the parameters that the United States saw as appropriate.

Ultimately, this is the crucial point in evaluating US policy during the operation and its significance for the future relations between the two countries.

Notes

Introduction

1   Israel Katz, "How should Israel bring an end to its responsibility for Gaza," *Makor Rishon*, April 1, 2018,
    https://www.makorrishon.co.il/opinion/34293/

2   Zaki Shalom, Israel, "The United States, And The Battle Over The Settlement Construction Freeze," 2009–2010. Institute for National Security Studies (Hereafter: INSS), 2017, p. 25. (Hereafter: Shalom, "The Battle Over The Settlement Construction Freeze").

3   Ibid., pp. 69–70.

4   Ibid.

5   Sputnik News Staff, "Lavrov Reveals Why Russia Turned Down Obama's 2016 Palestine Proposition," *Sputnik News*, September 24, 2018,
    https://sputniknews.com/world/201809241068284638-israel-palestine-conflict-obama-proposal/

6   Katz, Yaakov, "Why Israel has the most technologically advanced military on earth," *New York Post*, January 29, 2017,
    https://nypost.com/2017/01/29/why-israel-has-the-most-technologically-advanced-military-on-earth/

7   Yotam Berger, "Jewish settlement in the West Bank – Historical Perspective (in Hebrew)," *Haaretz*, June 6, 2017,
    https://www.haaretz.co.il/news/50years/.premium-MAGAZINE-1.4159019

8   Ibid.

9   Moshe Lisak, "The effects of the Intifada on the Israeli society," In Reuven Gal (Editor): *The Seventh War: The Effects of the Intifada on the Israeli Society*, Hakibutz Hmeuchad publications, Raanana, 1990, pp. 8–17.

10 Israel Intelligence Heritage Commemoration Center (Hereafter: IICC), "The suicide Palestinian terror against Israel: September 2000-December 2005," January 1, 2006,
    https://www.terrorism-info.org.il/Data/pdf/PDF_18891_1.pdf. See also: Maj. Gen. (Res.) Sami Turjeman, IDF, "The Road to Protective Edge," Policy Notes: The Washington Institute for Near East Policy 47, (2018): p. 2. (Hereafter: Turjeman, "The Road To Protective Edge").

11 IICC Newsletter, April 10, 2002,
    https://www.terrorisminfo.org.il/app/uploads/2017/09/6880018h-2.pdf

12 Zaki Shalom, Yoaz Hendel, "The Unique Features of the Second Intifada," *Military and Strategic Affairs*, Volume 3, No. 1, May 2011,
    http://www.inss.org.il/publication/the-unique-features-of-the-second-

intifada/ (Hereafter: Shalom and Hendel, The Unique Features of the Second Intifada).

13   *Davar* [Israeli Newspaper], February 26, 1996.

14   IICC Newsletter, "The Revolving Door Practice in the Palestinian Entity During Arafat Rule," October 23, 2007, https://www.terrorisminfo.org.il/Data/pdf/PDF_07_223_1.pdf.il/Data/pdf/PDF_07_223_1.pdf

15   Shalom and Hendel, "The Unique Features of the Second Intifada."

16   Ibid.

17   Ze'ev Drori, Dan Shomron,*Subtle Leadership*, Yediot Ahronot Publications, Tel Aviv, 2016, p. 366.

18   Neri Livneh, "The IDF must change its patterns of war against the Palestinians," *Haaretz*, December 8, 2007.

19   In September 2003, 27 pilots of the Israeli Air Force published a letter they sent to the Chief of Staff informing him that they refuse to take part in air strikes carried out in populated areas. These strikes they said were both immoral and illegal. This letter was written following the killing of a Hamas leader Saleh Shehada in Gaza on July 23, 2002, In this attack 15 civilian people including children were killed. See Amira Hess. "The pilots' letter following the killing of Saleh Shehada," *Haaretz*, January 7, 2011, https://www.haaretz.co.il/misc/1.1155643

20   Alexander Yaakobson, "The narrative of colonial struggle in the Israeli-Palestinian conflict," *Haaretz*, October 18, 2018.

21   Shmuel Even, "Abu Mazen's Opposition to Recognition of Israel as a Jewish State: Strategic Implications," http://www.inss.org.il/publication/abu-mazens-opposition-to-recognition-of-israel-as-a-jewish-state-strategic-implications/ (see also: Yonatan Lees, "There Will Be No Agreement Without Palestinian Recognition of Israel as a Jewish State," *Haaretz*, February 3, 2014, https://www.haaretz.co.il/news/politics/1.2234194

22   Zaki Shalom and Yoaz Hendel, *Defeating Terror: The Story Behind Israel's Victory over the Palestinian Intifada*, Yediot Ahronot Publications, Tel Aviv, 2010, p. 65.

23   Amos Yadlin, Udi Dekel Kim Lavi, "A Srategic Framework for the Israeli-Palestinian Arena," INSS, August 2018, p. 20.

24   Ehud Barak Testimony to the Vinograd Committee, November 28, 2006, https://www.makorrishon.co.il/nrg/images/news1/Ehud_Barak%202006.pdf

25   Mark A. Heller, "Implications of the Withdrawal from Lebanon for Israeli–Palestinian Relations," Strategic Assessment, Volume 3, No. 1, INSS, June 2000, http://www.inss.org.il/publication/implications-of-the-withdrawal-from-lebanon-for-israeli-palestinian-relations/

26   Zaki Shalom, "The Disengagement Plan: Vision and Reality," INSS, Strategic Assessment, Volume 13, No. 3, October 2010,

http://www.inss.org.il/publication/the-disengagement-plan-vision-and-reality/

27 Turjeman, "The Road to Protective Edge," p. 4.

28 Shmuel Even, "Israel's Strategy of Unilateral Withdrawal," INSS, Strategic Assessment, Volume 12, No. 1, June 2009, http://www.inss.org.il/publication/israels-strategy-of-unilateral-withdrawal/?offset=0&posts=1&outher=Shmuel%20Even&from_date=2009&to_date=2009&free_text=GAZA

29 Shalom, "The Battle Over The Settlement Construction Freeze, 2009–2010," pp. 71–89.

30 Prime Minister Netanyahu meeting with Italian President, Sergio Mattarella, November 2, 2016, Prime Minister's Office, https://www.gov.il/he/Departments/news/eventitaly021116

31 Shalom, "The Battle Over The Settlement Construction Freeze, 2009–2010," pp. 71–89.

32 The Reut Institute, "A Demilitarized Palestinian State," July 20, 2009, http://reut-institute.org/he/Publication.aspx?PublicationId=3674 (in Hebrew).

33 Shalom, "The Battle Over The Settlement Construction Freeze, 2009–2010," pp. 71–89.

34 Ibid.

35 Ibid.

36 Ruairi Casey, "Palestinian President Mahmoud Abbas: 'Jerusalem is not for sale,'" *Al-Jazeera*, September 28, 2018, https://www.aljazeera.com/news/2018/09/palestinian-president-mahmoud-abbas-jerusalem-sale-180927185800909.html

37 Shalom, "The Battle Over The Settlement Construction Freeze, 2009–2010," pp. 71–89.

38 Arafat's Johannesburg Speech, May 10, 1994, https://iris.org.il/quotes/joburg.htm

39 Full Speech by Hezbollah Secretary General, His Eminence Sayed Hassan Nasrallah, during the Bent Jbeil Victory Festival on May 26, 2000, http://www.english.alahednews.com.lb/essaydetails.php?eid=14178&cid=359#.WPMXWGdByUl

40 Turjeman, "The Road to Protective Edge," p. 2.

41 Shalom and Hendel, "The Unique Features of the Second Intifada," *Military and Strategic Affairs*, Volume 3, No. 1, May 2011.

42 During Cast Lead Operation the IDF prevented the entrance of journalists into the areas of battle; all the reports of journalists were strictly examined by the censorship. This did really help and detailed reports found their way into written and electronic media and especially to the social media. Nachman Shai, *Media War – Reaching for the Hearts and Minds*, Yediot Ahronot Publications, Tel Aviv 2013, p. 156.

1 Hamas and Israel

1 Joshua Mitnick, "Israel, Hamas Escalate Violence," *Wall Street Journal*, July 9, 2014,
https://www.wsj.com/articles/israel-steps-up-airstrikes-in-the-Gaza-strip-1404823112 (Hereafter: Mitnick, July 9, 2014)

2 Hamas Covenant 1988, "The Covenant of the Islamic Resistance Movement," August 18, 1988,
http://avalon.law.yale.edu/20th_century/hamas.asp (Hereafter: Hamas Covenant).

3 On the history of the Hamas see: Shaul Mishal and Avraham Sela, "The Hamas a Behavioral Profile," The Tami Steinmetz Center for Peace Research, Tel Aviv University, 1997.

4 Ibid.

5 Hamas Covenant 1988.

6 Ibid.

7 Jim Zanotti, "Hamas: Background and Issues for Congress," Congressional Research Center, December 2, 2010,
https://fas.org/sgp/crs/mideast/ R41514.pdf (Hereafter: Jim Zanotti, Hamas Background), p. 1.

8 IICC Newsletter, December 22, 2014.

9 IICC Newsletter, December 25, 2014.

10 Jim Zanotti, Hamas Background, p. 6.

11 *BBC News*, "Hamas Declares Israel Truce Over," December 22, 2008,
http://news.bbc.co.uk/2/hi/middle_east/7791100.stm

12 Daoud Kuttab, "Has Israel's Gaza Attack Revived Hamas?" *Washington Post*, December 30, 2008,
http://www.washingtonpost.com/wp-dyn/content/article/2008/12/29/AR2008122901901.html

13 Prime Minister Ehud Olmert's Speech at the Institute for National Security Studies Annual Conference. Prime Minister's Office, December 18, 2008,
https://www.gov.il/he/Departments/news/speechdef181208

14 Jim Zanotti, "Israel and Hamas: Conflict in Gaza (2008–2009), Congressional Research Center, February 19, 2009,
https://fas.org/sgp/crs/mideast/R40101.pdf (Hereafter: Jim Zanotti, Israel and Hamas), p. 7.

15 Israeli Security Service (Shabak), "One year to Operation Cast Lead,"
https://www.shabak.gov.il/publications/Pages/study/oferetsummary.aspx

16 IICC Newsletter, "A survey of the rockets strikes against Israel by the Hamas in 2008," January 1, 2009.

17 IICC Newsletter, December 28, 2014. See also: Has Israel's Gaza Attack Revived Hamas?" *Washington Post*, December 30, 2008.

18 Prime Minister Ehud Olmert's Remarks at the Press Briefing
On the Operation in the Gaza Strip, Prime Minister's Office, 27/12/2008
https://www.gov.il/he/Departments/news/speechGaza271208

19 A survey of the Israel Democracy Institute, January 4–6, 2009,
http://www.peaceindex.org/files/peaceindex2008_12_9.pdf

20 Prime Minister Ehud Olmert's Remarks at the Press Briefing

On the Operation in the Gaza Strip, Prime Minister's Office, December 27, 2008,
https://www.gov.il/he/Departments/news/speechGaza271208

21 Ibid.

22 Jim Zanotti, Israel and Hamas, pp. 7–8.

23 IICC Newsletter, December 29, 2008.

24 IICC Newsletter, December 31, 2008.

25 Ibid.

26 Prime Minister's speech at air force base Hatzerim, Prime Minister's Office, December 30, 2008,
https://www.gov.il/he/Departments/news/ eventsouth301208. See also: IICC Newsletter, December 30, 2008.

27 IICC Newsletter, January 5, 2009.

28 IICC Newsletter, January 6, 2009.

29 As quoted in Josh Reubner, "Shattered Hopes: Obama's Failure To Broker Israeli–Palestinian Peace" (2013): p. 44.

30 Jewish Virtual Library, "American Public Opinion Polls: Opinion Toward Israeli Operations in Gaza (2009–2014),"
http://www.jewishvirtuallibrary.org/american-opinion-toward-israeli-operations-quot-cast-lead-quot-and-quot-pillar-of-defense-quot

31 Ibid.

32 Prime Minister Olmert meeting with NATO Secretary Jaap De Hoop Shepherd, Prime Minister's Office, January 11, 2009,
https://www.gov.il/he/Departments/news/eventnato110109

33 Gabi Siboni, "Operations Cast Lead, Pillar of Defense, and Protective Edge: A Comparative Review," In: The Lessons of Protective Edge (Anat Kurz and Shlomo Brom, Editors), 2014, pp. 28–29.
http://www.inss.org.il/wp-content/uploads/systemfiles/ Zuk EtanENG final.pdf (Hereafter: Siboni, "Operations Cast Lead, Pillar of Defense, and Protective Edge").

34 Jim Zanotti, Israel and Hamas, p. 8.

35 IICC Newsletter, January 13, 2009.

36 IICC Newsletter, January 19, 2009.

37 Ibid.

38 Prime Minister Olmert's speech following a cabinet, meeting, Prime Minister's Office, January 17, 2009, https://www.gov.il/he/Departments/ news/speechcabinet170109. (In Hebrew).
The issue of "moral fighting" is one of the most controversial issues within the Israeli society. There are many who claim that the IDF is conducting a policy that puts the life of Israeli soldiers in danger in order to prevent the killing of innocent civilians. Such views are mainly adopted by the religious right-wing. They claim that according to Jewish heritage whenever there is a danger to an Israeli soldier and a civilian Palestinian the life of the soldier should be given priority over the lives of others. Levy Yagil, The Divine Commander: The Theocratization of the Israeli. Military, Am Oved Publishing, Tel Aviv, 2015, p. 340.

39 Oded Eran, "Operation Cast Lead: The Diplomatic Dimension," INSS,

Strategic Assessment, volume 11, February 2009. See also: Hirsh Goodman, "Israel's Public Diplomacy in Operation Cast Lead," INSS Insight No. 90, January 15, 2009, http://www.inss.org.il/publication/israels-public-diplomacy-in-operation-cast-lead/

40   Prime Minister Ehud Olmert's Statement at the Summit With European Leaders, January 18, 2009, https://www.gov.il/he/Departments/news/speechleaders180109

41   Ibid.

42   Ibid.

43   Ibid.

44   Ibid.

45   Zaki Shalom, "The War against the Hamas in Cast Lead Operation," INSS, Strategic Assessment, volume 4, February 2009, pp. 69—2.

46   Ephraim Kam, "The Arab Reaction to Operation Cast Lead," INSS Insight No. 86, January 6, 2009, http://www.inss.org.il/publication/the-arab-reaction-to-operation-cast-lead/

47   Giora Eiland, "Operation Pillar of Defense: Strategic Perspectives," In: In the Aftermath of Operation Pillar of Defense, The Gaza Strip, November 2012 (Shlomo Brom, Editor), December 2012, p. 12. http://www.inss.org.il/uploadImages/systemFiles/memo124f027134590.pdf

48   A survey conducted by the Israel Democracy Institute, November 28, 2012, http://www.peaceindex.org/files/... pdf

49   IICC Newsletter, November 15, 2012, https://www.terrorism-info.org.il/he/20425/

50   Ehud Yaari, "Hamas: Israel opened the Gates of hell," *Mako News*, November 14, 2012. https://www.mako.co.il/news-military/security/Article-1ba679e927ffa31006.htm

51   Prime Minister Netanyahu meeting with the Secretary of the United Nations, November 20, 2012, Prime Minister's Office, https://www.gov.il/he/Departments/news/eventban201112

52   Prime Minister Netanyahu meeting German foreign minister, November 20, 2012. Prime Minister's Office, https://www.gov.il/he/Departments/news/eventgermany201112

53   Netanyahu speech at the Ben-Gurion ceremony, November 20, 2012. Prime Minister's Office, https://www.gov.il/he/Departments/news/speechbengurion201112

54   IICC Newsletter, November 21, 2012. Newsletter. https://www.terrorism-info.org.il/he/20671/ (Hereafter: IICC Newsletter, November 21, 2012).

55   IICC Newsletter, November 22, 2012, newsletter. https://www.terrorism-info.org.il/he/20433 / (Hereafter: IICC Newsletter, November 22, 2012)

56   Avner Golov, "The Campaign to Restore Israeli Deterrence," In: In the Aftermath of Operation Pillar of Defense, The Gaza Strip, INSS, November 2012, (Shlomo Brom, Editor), December 2012, p. 26. http://www.inss.org.il/uploadImages/systemFiles/memo124f027134590.pdf

57 Ibid.
58 Prime Minister Netanyahu's statement at the end of Operation Pillar of Defense, Prime Minister's Office, November 21, 2012, https://www.gov.il/he/Departments/news/eventsumm211112. See also: Oded Eran, "International Aspects of Operation Pillar of Defense," INSS Insight No. 386, November 21, 2012, http://www.inss.org.il/publication/international-aspects-of-operation-pillar-of-defense/
59 IICC Newsletter, November 21, 2012.
60 IICC Newsletter, November 22, 2012.
61 Siboni, "Operations Cast Lead, Pillar of Defense, and Protective Edge," p. 29.

2 Operation Protective Edge

1   Kay Armin Serjoie, "One Result of the Gaza Conflict: Iran and Hamas Are Back Together," *Time*, August 19, 2014, http://time.com/3138366/iran-and-hamas-alliance-after-Gaza-war/
2   Nidal al-Mughrabi, "Hamas' deputy chief says it has patched up ties with Iran," *Reuters*, December 17, 2014. http://www.reuters.com/article/us-mideast-hamas-Gaza-idUSKBN0JV1NH2014121 7
3   Siboni, "Operations Cast Lead, Pillar of Defense, and Protective Edge," p. 29.
4   "Gaza conflict: Can economic isolation ever be reversed?" *CNN*, August 4, 2014, http://edition.cnn.com/2014/08/03/business/Gaza-economic-isolation-mme/ (Hereafter: "Gaza Conflict: Can economic isolation ever be reversed?").
5   Sultan Barakat and Omar Shaban, "Back To Gaza: A New Approach to Reconstruction," *Brookings Policy Briefing* (2015): p. 6. (Hereafter: Barakat and Shaban, Back To Gaza).
6   "Egypt court puts Hamas on terrorist list," *BBC*, 28 February 2015, https://www.bbc.com/news/world-middle-east-31674458
7   Hirsh Goodman, "Israel's Narrative," *Jerusalem Center for Public Affairs*, In *The Gaza War 2014: The War Israel Did Not Want and the Disaster It Averted* (Hirsh Goodman and Dore Gold, editors), p. 11. (Hereafter: Goodman, Israel's Narrative, 2014).
8   Barakat and Shaban, Back To Gaza, p. 6.
9   "Gaza Conflict: Can economic isolation ever be reversed?"
10 Dr Jeroen Gunning Executive Director, Durham Global Security Institute, What drove Hamas to take on Israel? 18 July 2014, http://www.bbc.com/news/world-middle-east-28371966
11 Ibid.
12 Israel Defense Forces, https://www.idf.il/en/minisites/hamas/the-2014-kidnapping-of-3-israeli-teens/
13 TOI staff, "Obama: I cannot imagine indescribable pain of teens' parents," June 30, 2014, *Times of Israel*, https://www.timesofisrael.com/international-leaders-condemn-murder-of-kidnapped-teens/

14 Prime Minister Netanyahu's speech at the Prime Minister's Office, June 14, 2014, https://www.gov.il/he/Departments/news/spokestate140614

15 Prime Minister Netanyahu's speech to the foreign press, June 15, 2014, https://www.gov.il/he/Departments/news/speechstatement150614

16 Secretary of State John Kerry, Press Statement, June 15, 2014, https://2009-2017.state.gov/secretary/remarks/2014/06/227599.htm

17 Ibid.

18 Ibid.

19 Prime Minister Netanyahu's speech, June 16, 2014, Prime Minister's Office, https://www.gov.il/he/Departments/news/eventcon160614

20 Prime Minister Netanyahu's meeting with Tony Blair, June 17, 2014. Prime Minister's Office, https://www.gov.il/he/Departments/news/eventblair170614

21 Prime Minister Netanyahu's meeting with mayors, June 18, 2014, Prime Minister's Office, https://www.gov.il/he/Departments/news/mayorsevent180614

22 MFA Behind the Headlines: The Hamas Kidnapping of Three Israeli Teens, http://mfa.gov.il/MFA/ForeignPolicy/Issues/Pages/Behind-the-Headlines-The-Hamas-kidnapping-of-three-Israeli-teens-15-Jun-2014.aspx

23 Prime Minister Netanyahu's speech in Judea and Samaria, June 19, 2014, Prime Minister's Office, https://www.gov.il/he/Departments/news/eventhatmar190614

24 Prime Minister Netanyahu's meeting with the parents of the three boys, June 19, 2014, Prime Minister's Office, https://www.gov.il/he/Departments/news/eventfam200614
See also Prime Minister Netanyahu's speech at the Jewish Agency, November 23, 2014, Prime Minister's Office, https://www.gov.il/he/Departments/news/speechsochnut230614

25 Prime Minister's Office, June 26, 2014, https://www.gov.il/he/Departments/news/speechwings260614

26 Elhanan Miller, "Hamas chief claims 3 kidnapped youths were soldiers, Khaled Mashal claims no responsibility for kidnapping but blesses the perpetrators, blames Netanyahu for 'Judaizing' Jerusalem," *Times of Israel*, June 24, 2014, https://www.timesofisrael.com/hamas-chief-claims-no-responsibility-for-kidnapping-but-praises-it/

27 Prime Minister's Office, June 29, 2014, https://www.gov.il/he/Departments/news/speechinss290614

28 Ibid.

29 Prime Minister's Office, July 6, 2014, https://www.gov.il/he/Departments/news/eventyifrach060714

30 Prime Minister Netanyahu speech at the funeral of the three boys, July 1, 2014, Prime Minister Office, https://www.gov.il/he/Departments/news/eventeulogy010714

31 Jodi Rudoren and Isabel Kershner, "Israel's Search for 3 Teenagers Ends in Grief," *The New York Times*, June 30, 2014,

https://www.nytimes.com/2014/07/01/world/middleeast/Israel-missing-teenagers.html

32 Barak Ravid, "Netanyahu on Murders of Three Israeli Teens: Hamas Is Responsible and Hamas Will Pay," *Haaretz*, June 30, 2014, https://www.haaretz.com/.premium-pm-hamas-is-responsible-will-pay-1.5253993

33 Statement by the President on the Deaths of Naftali Fraenkel, Eyal Yifrach, and Gilad Shaar, The White House, Office of the Press Secretary, June 30, 2014, https://obamawhitehouse.archives.gov/the-press-office/2014/06/30/statement-president-deaths-naftali-fraenkel-eyal-yifrach-and-gilad-shaar

34 Ibid.

35 Al-Jazeera, "Shocked Palestinian Family Waits to Bury Son."

36 Middle East Media Research Institute (Hereafter: MEMRI), July 4, 2014, http://www.memri.org.il/cgi-webaxy/item?3664&findWords= [...]

37 Haviv Rettig Gur, "No Place in Israel for Abu Khdeir's killers, Netanyahu says," *Times of Israel*, July 6 2014, https://www.timesofisrael.com/no-place-in-israel-for-abu-khdeirs-killers-netanyahu-says/

38 Ibid.

39 Jack Khoury, Barak Ravid, Jonathan Lis, "Abbas Blames Israel for Teen's Murder, Demands Netanyahu Condemn it," *Haaretz*, July 3, 2014, https://www.haaretz.com/.premium-abbas-demands-netanyahu-condemn-teen-s-murder-1.5254280

40 Ibid.

41 *Jerusalem Post*, "Condemnations From Around the World Poring over Murder of Palestinian Youth."

42 Joshua Mitnick, Tamer El-Ghobashy, and Sara Toth Stub, "Gaza Rockets Reaching Deeper Into Israel," *The Wall Street Journal*, July 10, https://www.wsj.com/articles/israel-and-palestinian-militants-exchange-fire-as-confrontation-continues-1404908708

43 Database Desk, "Operation 'Protective Edge': A Detailed Summary of Events," *International Institute for Counter-Terrorism*, July 12, 2014, https://www.ict.org.il/Article/1262/Operation-Protective-Edge-A-Detailed-Summary-of-Events (Hereafter: Operation 'Protective Edge': A Detailed Summary of Events).

44 Mitnick, July 9, 2014.

45 Rubenstein Daniel, "Hamas' Tunnel Network: A Massacre in the Making," *Jerusalem Center for Public Affairs* (Hereafter: JCPA), February 26, 2015, http://jcpa.org/hamas-tunnel-network/

46 Goodman, Israel's Narrative, 2014, p. 8.

47 Israeli Prime Minister's Office, July 7, 2014, https://www.gov.il/he/Departments/news/spokestart060714

48 A Survey of Israel Democracy Institute, July 23, 2014, http://www.peaceindex.org/files/Peace_Index_Data_July_2014-Heb.pdf

49 Israeli Prime Minister's Office, July 8, 2014, https://www.gov.il/he/Departments/news/spokebibi080714

50   Prime Minister Netanyahu's visit in the southern command, July 9, 2014, Prime Minister's Office, https://www.gov.il/he/Departments/news/eventsoutherncommand 090714

51   Ibid.

52   Israeli Prime Minister's Office, July 10, 2014, https://www.gov.il/he/Departments/news/spokecabinet100714

53   IICC Newsletter, July 13, 2014, https://www.terrorism-info.org.il/he/20670/

54   Israeli Prime Minister's Office, July 11, 2014, https://www.gov.il/he/Departments/news/eventstate110714

55   Ibid.

56   Israeli Prime Minister's Office, July 13, 2014, https://www.gov.il/he/Departments/news/eventinterview130714

57   Ibid.

58   Israeli Prime Minister's Office, July 15, 2014, https://www.gov.il/he/Departments/news/eventpress150714

59   Israeli Prime Minister's Office, July 16, 2014, https://www.gov.il/he/Departments/news/spokemeetita160714

60   Israeli Prime Minister Office, July 22, 2014, https://www.gov.il/he/Departments/news/eventkimoon220714

61   Ibid.

62   Ibid.

63   Ibid.

64   Ibid.

65   Prime Minister's Office, July 20, 2014, https://www.gov.il/he/Departments/news/speechkirya200714

66   Ibid.

67   Ibid.

68   Prime Minister's Office, July 24, 2014, https://www.gov.il/he/Departments/news/eventbritishfm240714

69   Ibid.

70   Prime Minister's Office, July 28, 2014, https://www.gov.il/he/Departments/news/eventkirya280714

71   Prime Minister's Office, August 28, 2014, https://www.gov.il/he/Departments/news/eventhoward280814

72   Prime Minister's Office, August 27, 2014, https://www.gov.il/he/Departments/news/speechstat270814 See: Yiftah Shapir, "How Many Rockets Did Iron Dome Shoot Down?" INSS Insight No. 414, March 21, 2013.

73   Ibid.

74   IICC Newsletter, September 2, 2014. https://www.terrorism-info.org.il/he/20709/

75   Pnina Sharvit Baruch, "Operation Protective Edge: Legality and Legitimacy," INSS, July 22, 2014, http://www.inss.org.il/publication/operation-protective-edge-legality-and-legitimacy/

76 Ibid.

77 State Comptroller of Israel, Operation "Protective Edge" – IDF Activity from the Perspective of International Law, Particularly with Regard to Mechanisms of Examination and Oversight of Civilian and Military Echelons, p. 10, http://www.mevaker.gov.il/he/Reports/Pages/622.aspx?AspxAuto DetectCookieSupport=1 (Hereafter: State Comptroller of Israel, Operation "Protective Edge").

78 Ibid.

79 Operation 'Protective Edge': A Detailed Summary of Events.

80 Ibid.

81 Israel Ministry of Foreign Affairs, "The 2014 Gaza Conflict 7 July–26 August 2014: Factual and Legal Aspects," May 2015, p. xi. (Hereafter: Israel Ministry of Foreign Affairs, May 2015).

82 State Comptroller of Israel, Operation "Protective Edge," p. 29.

83 Ministry of Foreign Affairs, May 2015, p. 30.

84 State Comptroller of Israel, Operation "Protective Edge," p. 163.

85 Ibid.

86 Lt. Col. (res) David Benjamin, "Israel, Gaza and Humanitarian Law: Efforts to Limit Civilian Casualties), *Jerusalem Center for Public Affairs*, http://jcpa.org/the-gaza-war-2014/israel-gaza-humanitarian-law-civilian-casualties/ (Hereafter: David Benjamin, Israel, Gaza and Humanitarian Law).

87 Richard Goldstone, "Reconsidering the Goldstone Report on Israel and war crimes," *Washington Post*, April 1, 2011, https://www.washingtonpost.com/opinions/reconsidering-the-goldstone-report-on-israel-and-war-crimes/2011/04/01/AFg111JC story.html?utm_term=.ed616cdfba7d

88 David Benjamin, Israel, Gaza and Humanitarian Law.

89 Ministry of Foreign Affairs, May 2015, p. 97.

90 Mitnick, July 9, 2014.

91 Ministry of Foreign Affairs, May 2015 p. 48.

92 Ministry of Foreign Affairs, May 2015, p. 156.

93 Ministry of Foreign Affairs, May 2015, p. 179.

94 Ministry of Foreign Affairs, May 2015, p. 200.

95 Ministry of Foreign Affairs, May 2015, p. 202.

96 Ministry of Foreign Affairs, May 2015, p. 208.

97 Ibid.

98 Emily Landau and Azriel Bermant, "Iron Dome Protection: Missile Defense in Israel's Security Concept." Page. 41 in *The Lessons of Operation Protective Edge*, eds. Anat Kurz and Shlomo Brom, Tel Aviv: Institute for National Security Studies, 2014.

99 Thane Rosenbaum , "Hamas's Civilian Death Strategy Gazans shelter terrorists and their weapons in their homes, right beside sofas and dirty diapers," *Wall Street Journal*, July 21, 2014, https://www.wsj.com/articles/thane-rosenbaum-civilian-casualties-in-gaza-1405970362

100 Raphael S. Cohen, David E. Johnson, David E. Thaler, Brenna Allen, Elizabeth M. Bartels, James Cahill, Shira Efron, "From Cast Lead to Protective Edge," *Rand Corporation* (2017): xv.

101 Ministry of Foreign Affairs, May 2015, p. 74.

102 Turjeman, "The Road to Protective Edge," p. 5.

103 Ministry of Foreign Affairs, "Hamas terrorists confess to using human shields," August 27, 2014, http://mfa.gov.il/MFA/ForeignPolicy/Terrorism/Pages/Hamas-terrorists-confess-to-using-human-shields.aspx

104 Ministry of Foreign Affairs, May 2015, p. 84.

105 Ministry of Foreign Affairs, May 2015, p. 158.

106 Ministry of Foreign Affairs, "IDF report: Hamas illegally used civilian infrastructure during Operation Protective Edge," August 21, 2014, http://mfa.gov.il/MFA/ForeignPolicy/Terrorism/Pages/IDF-report-Hamas-illegally-used-civilian-infrastructure-during-Operation-Protective-Edge.aspx

107 White Jeffrey, "Six Ways Hamas Could Limit Civilian Casualties in Gaza," *The Washington Institute*, August 23, 2014, http://www.washingtoninstitute.org/policy-analysis/view/six-ways-hamas-could-limit-civilian-casualties-in-Gaza

108 Ibid.

109 Ibid.

110 Ibid.

111 Pollock David, "How Many Civilians Have Been Killed in Gaza," *The Washington Institute*, August 20, 2014, http://www.washingtoninstitute.org/policy-analysis/view/how-many-civilians-have-been-killed-in-Gaza

112 Ibid.

113 Ibid.

114 Ministry of Foreign Affairs, May 2015, viii.

115 Ben-David Lenny, "Gaza Casualties: How Many and Who They Were," *Jerusalem Center for Public Affairs*, February 26, 2015, http://jcpa.org/casualties-Gaza-war/

116 Steven Stotsky, "How Hamas Wields Gaza's Casualties as Propaganda, *Time*, July 29, 2014, http://time.com/3035937/Gaza-israel-hamas-palestinian-casualties/

117 Ibid.

118 Goodman, Israel's Narrative, 2014, p. 17.

119 Ministry of Foreign Affairs, May 2015, p. 2.

120 Raphael S. Cohen, David E. Johnson, David E. Thaler, Brenna Allen, Elizabeth M. Bartels, James Cahill, Shira Efron, "From Cast Lead to Protective Edge," *Rand Corporation* (2017): p. 83. (Hereafter: Rand Corporation).

121 Rand Corporation, p. 84.

122 Rand Corporation, p. 85.

123 Rand Corporation, p. 86.

124 Rand Corporation, p. 87.

125 Rand Corporation, p. 88.

126 Ibid.

127 Rand Corporation, p. 91.

128 Ibid.

129 Rand Corporation, pp. 93–94.

130 Rand Corporation, p. 95.

131 Rand Corporation, p. 97.

132 Udi Dekel and Shlomo Brom, "The Second Stage of Operation Protective Edge: A Limited Ground Maneuver," *INSS*, July 21, 2014, http://www.inss.org.il/publication/the-second-stage-of-operation-protective-edge-a-limited-ground-maneuver/

133 Ibid.

134 Rand Corporation, p. 99.

135 Rand Corporation, p. 97.

136 Rand Corporation, p. 101.

137 Rand Corporation, p. 102.

138 Rand Corporation, pp. 102–103.

139 Rand Corporation, p. 105.

140 Ibid.

141 Rand Corporation, p. 107.

142 Ibid.

143 Rand Corporation pp. 104–108.

144 Rand Corporation, p. 108.

145 Rand Corporation, p. 110.

146 Ibid.

147 Rand Corporation, pp. 110–111.

148 Rand Corporation, p. 112.

149 Rand Corporation, p. 114.

150 Rand Corporation, p. 115.

151 Rand Corporation, p. 117.

152 Ibid.

153 Ibid.

154 Ibid.

155 Rand Corporation, p. 118.

156 Rand Corporation, p. 117.

157 Rand Corporation, p. 118.

158 Rand Corporation, pp. 118–120.

159 Rand Corporation, p. 120.

160 Ibid.

161 Ibid.

162 Rand Corporation, p. 122.

163 Rand Corporation, p. 124.

164 Rand Corporation, p. 125.

165 Ibid.

166 Ibid.

167 Rand Corporation, p. 124.

168 Rand Corporation, p. 125.

169 Rand Corporation, p. 126.

170 Ibid.

171 Meir Elran, Yehuda Ben Meir, Gilead Sher, "The Impact of Operation Protective Edge on Political and Social Trends in Israel," in Strategic Survey for Israel 2014–2015, eds. Anat Kurz and Shlomo Brom, Tel Aviv: Institute for National Security Studies, 2015, p. 149, http://www.inss.org.il/wp-content/uploads/systemfiles/ INSS2014-15Balance_ENG%20(2)_Elran,%20Ben%20Meir, %20Sher.pdf

172 Ibid., p. 147.

173 Rand Corporation, pp. 126–127.

174 Rand Corporation, p. 127.

175 Ibid.

176 Maj. Gen. (res.) Yaakov Amidror, "Misplaced Frustration," *Begin–Sadat Center for Strategic Studies*, September 3, 2014, https://besacenter.org/perspectives-papers/misplaced-frustration/

177 MEMRI: Hamas: Abduction Of Three Israeli Youths Marks Start Of New Intifada In West Bank, End Of Palestinian Authority; July 18, 2014; https://www.memri.org/reports/hamas-abduction-three-israeli-youths- marks-start-new-intifada-west-bank-end-palestinian

178 MEMRI: The abduction of the teenagers, the Hamas reaction, June 19, 2014, http://www.memri.org.il/cgi-webaxy/item?3655&findWords=[...]

179 MEMRI: Hamas: Abduction Of Three Israeli Youths Marks Start Of New Intifada In West Bank, End Of Palestinian Authority; July 18, 2014, https://www.memri.org/reports/hamas-abduction-three-israeli-youths- marks-start-new-intifada-west-bank-end-palestinian

180 MEMRI: The abduction of the teenagers, the Hamas reaction, June 19, 2014, http://www.memri.org.il/cgi-webaxy/item?3655&findWords= [...]

181 MEMRI: PA Chairman Mahmoud Abbas Defends Security Coordination with Israel, Denounces Abduction of Israeli Youth; June 18, 2014, https://www.memri.org/tv/pa-chairman-mahmoud-abbas-defends- security-coordination-israel-denounces-abduction-israeli-youth /transcript

182 Ibid.

183 MEMRI: PA and Fatah Positions on the Kidnapping of the Three Israeli Teens; July 25, 2014, https://www.memri.org/reports/pa-and-fatah-positions-kidnapping- three-israeli-teens

184 Ibid.

185 Ibid.

186 MEMRI: Hamas, Palestinian Authority Trade Accusations Over Israeli Operation in Gaza; July 14, 2, https://www.memri.org/reports/hamas-palestinian-authority-trade- accusations-over-israeli-operation-gaza

187 Ibid.

188 Ibid.

189 Ibid.
190 MEMRI: Palestinian Leadership U-Turn on Gaza Conflict; July 23, 2014, https://www.memri.org/reports/palestinian-leadership-u-turn-gaza-conflict-attacking-israel-adopting-hamas-conditions
191 Ibid.
192 MEMRI: Fatah Official Makes Impassioned Plea for Immediate Ceasefire; July 23, 2014, https://www.memri.org/tv/fatah-official-makes-impassioned-plea-immediate-ceasefire-stopping-bloodshed-our-number-one/transcript
193 MEMRI: PLO and Fatah; July 28, 2014, https://www.memri.org/reports/plo-and-fatah-kerry-initiative-paris-conference-%E2%80%93-plot-bypass-egypt-and-remove-plo
194 Ibid.
195 Ibid.
196 Ibid.
197 Ibid.
198 MEMRI: Jibril Rajoub Calls on Hamas to Disengage from the Muslim Brotherhood; August 13, 2014, https://www.memri.org/tv/jibril-rajoub-calls-hamas-disengage-muslim-brotherhood/transcript

3 Policy of the United States throughout Protective Edge

1 Prime Minister Ariel Sharon's Statement on the Day of the Implementation of the Disengagement Plan, August 15, 2005, https://www.gov.il/he/Departments/news/speech150805
2 Dore Gold, "Telling the Truth About the 2014 Gaza War," *Jerusalem Center for Public Affairs*, in *The Gaza War 2014: The War Israel Did Not Want and the Disaster It Averted* (Hirsh Goodman and Dore Gold, editors), p. 31. (Hereafter: Gold, "The Gaza War, 2014").
3 Goodman, Israel's Narrative, 2014, p. 7.
4 Gold, "The Gaza War, 2014," p. 31.
5 Prime Minister Olmert's speech, June 26, 2007, Prime Minister's Office, https://www.gov.il/he/Departments/news/speechkesarya280607
6 Gold, "The Gaza War, 2014," p. 33.
7 Gold, "The Gaza War, 2014," p. 34.
8 Gold, "The Gaza War, 2014," p. 35.
9 Gold, "The Gaza War, 2014," p. 36.
10 John Kerry remarks, July 21, 2014, https://2009-2017.state.gov/secretary/remarks/2014/07/229574.htm
11 "Kirk, Cardin, Rubio Demand Investigation into Weapons Found at UNRWA Facilities: Accountability for Hamas Rockets Discovered on UNRWA Premises; United States Is the Largest Single Donor to UNRWA," Wednesday, August 6, 2014, http://www.kirk.senate.gov/?p=press_release&id=1170.
12 Jen Psaki, State Department Spokesperson, Statement, "UNRWA School Shelling," Washington, DC, August 3,

http://www.state.gov/r/pa/prs/ps/2014/230160.htm

13 Ibid.

14 David Benjamin, Israel, Gaza and Humanitarian Law

15 Charles Krauthammer, "Moral Clarity in Gaza," *The Washington Post*, July 17, 2014,
https://www.washingtonpost.com/opinions/charles-krauthammer-moral-clarity-in-gaza/2014/07/17/0adabe0c-0de4-11e4-8c9a-923ecc0c7d23_story.html?utm_term=.6deded54274c

16 John Kerry, "Getting Caught Trying," in *Every Day is Extra*, 2018, p. 683. (Hereafter: Kerry, *Every Day is Extra*).

17 United States Department of State, Remarks With UN Secretary-General Ban Ki-moon, Egyptian Foreign Minister Sameh Shoukry, and Arab League Secretary General Nabil al-Araby, Remarks July 25, 2014,
https://2009-2017.state.gov/secretary/remarks/2014/07/229803.htm

18 As quoted in: Zaki Shalom, "The US Administration on Israel's Military Activity in Operation Protective Edge: Fluctuating Positions,"
http://www.inss.org.il/publication/the-us-administration-on-israels-military-activity-in-operation-protective-edge-fluctuating-positions/

19 Barak Ravid, "Kerry Condemns 'Brazen' Hamas Rockets, Urges Cease-fire," July 15, 2014,
http://www.haaretz.com/israelnews/1.605289?utm_source=dlvr.it&utm_medium=twitter

20 The White House, Office of the Press Secretary, Press Conference by the President, August 1, 2014,
https://obamawhitehouse.archives.gov/the-press-office/2014/08/01/press-conference-president

21 The Economic Times Staff, "US blames Hamas for collapse of Gaza cease-fire," *The Economic Times*, Aug 1, 2014,
http://economictimes.indiatimes.com/news/international/world-news/us-blames-hamas-for-collapse-of-Gaza-ceasefire/articleshow/39435364.cms

22 Ibid.

23 The White House, Office of the Press Secretary, Press Conference by the President, August 1, 2014,
https://obamawhitehouse.archives.gov/the-press-office/2014/08/01/press-conference-president

24 Rebecca Shimoni Stoli and AP, "Obama approves $225 million in Iron Dome funding," *Times of Israel*, August 5, 2014,
https://www.timesofisrael.com/obama-approves-225-million-in-iron-dome-funding/

25 PM Netanyahu's Statement at the Prime Minister's Office in Jerusalem, August 27, 2014,
http://www.pmo.gov.il/english/mediacenter/speeches/pages/speechstat270814.aspx.

26 The White House, Office of the Press Secretary, Press Conference by the President, August 1, 2014,
https://obamawhitehouse.archives.gov/the-press-office/2014/08/01/press-conference-president

27 Ibid.

28 Reuters Staff, "U.S. calls shelling outside U.N. school in Gaza 'disgraceful'," *Reuters*, August 3, 2014, https://www.reuters.com/article/us-mideast-gaza-school-usa/u-s-calls-shelling-outside-u-n-school-in-gaza-disgraceful-idUSKBN0G30NS2014 0803

29 Keith Ellison, "End the Gaza Blockade to Achieve Peace." *The Washington Post*, July 29, 2014, https://www.washingtonpost.com/opinions/keith-ellison-end-the-gaza-blockade-to-achieve-peace/2014/07/29/e5e707c4-16a1-11e4-85b6-c1451 e622637_story.html?utm_term=.cf16cc58e51d

30 Jimmy Carter and Mary Robinson, "How to Fix it." *Foreign Policy*, August 4, 2014, https://foreignpolicy.com/2014/08/04/how-to-fix-it/

31 Hamas Covenant.

32 John Kerry remarks with UN Secretary General Ban Ki-moon, July 21, 2014, https://2009-2017.state.gov/secretary/remarks/2014/07/229574.htm

33 Secretary of State John Kerry and Egyptian Foreign Minister Shoukry Joint Statements after Meeting with Egyptian President al-Sisi, July 22, 2014, https://2009-2017.state.gov/secretary/remarks/2014/07/229626.htm

34 Secretary of State John Kerry Secret Remarks with Turkish Foreign Minister Ahmet Davutoglu and Qatari Foreign Minister Khalid al-Attiyah after Their Meeting, July 26, 2014, http://www.state.gov/secretary/remarks/2014/07/229811.htm, (Hereafter: Kerry remarks, July 26, 2014).

35 *Every Day is Extra*, pp. 683–684.

36 Elliott Abrams, "What Now for Israel? Common Enemies and Shared Interests Have Aligned the Saudis, the Egyptians, and Other Arabs with the Jewish State. That's the Good News," *Mosaic*, September 2014, http://mosaicmagazine.com/essay/2014/09/what-now-for-israel/.

37 Raphael Ahren and AP, "Erdo an calls Israel more barbaric than Hitler", *Times of Israel*, July 19, 2014, http://www.timesofisrael.com/erdogan-calls-israel-more-barbaric-than-hitler/

38 AFP, "Erdo an says Israel attempting 'systematic genocide' in Gaza", *Times of Israel*, July 17, 2014, http://www.timesofisrael.com/erdogan-says-israel-atempting-systematic-genocide-in-Gaza/

39 Associated Press, "Turkey Declares 3-day Mourning Over Gaza," *Times of Israel*, July 21, 2014, http://www.timesofisrael.com/turkey-declares-3-day-mourning-over-Gaza/

40 Russia Today, "Israel Reduces Diplomatic Staff in Turkey Following Violent Pro-Palestinian Protests," *Russia Today*, July 18, 2014, https://www.rt.com/news/173992-turkey-protest-Gaza-israel/

41 Times of Israel Staff, "Kerry Under Fire in Israel for Negotiating with

Qatar, Turkey, not Israel, Egypt," *Times of Israel*, July 26, 2014, http://www.timesofisrael.com/kerry-under-fire-in-israel-for-negotiating-with-qatar-turkey-not-egypt-israel/

42 Kerry remarks, July 26, 2014.

43 Ibid.

44 Ibid.

45 Barak Ravid, "Kerry's Cease-fire Draft Revealed: U.S. Plan Would Let Hamas Keep Its Rockets," *Haaretz*, July 28, 2014, http://www.haaretz.com/israel-news/.premium-1.607379 (Hereafter: Barak Ravid, "Kerry's Cease-fire Draft Revealed").

46 Ibid.

47 Ibid.

48 Ibid.

49 David Horovitz, "John Kerry: The betrayal," *Times of Israel*, July 27, 2014, http://www.timesofisrael.com/john-kerry-the-betrayal/

50 Ibid.

51 Liveblog July 27, Day 20 of war with Hamas (closed), Haaretz: Kerry's proposal empowered Hamas, http://www.timesofisrael.com/liveblog_entry/haaretz-reporter kerrys-ceasefire-proposal-empowered-hamas/

52 Ibid.

53 MEMRI: PLO And Fatah: Kerry Initiative, Paris Conference – A Plot To Bypass Egypt And Remove The PLO, July 28, 2014, https://www.memri.org/reports/plo-and-fatah-kerry-initiative-paris-conference-plot-bypass-egypt-and-remove-plo (Hereafter: PLO And Fatah: "Kerry Initiative, Paris Conference – A Plot To Bypass Egypt And Remove The PLO").

54 Adam Entous, "Gaza Crisis: Israel Outflanks the White House on Strategy," *The Wall Street Journal*, August 14, 2014, https://www.wsj.com/articles/u-s-sway-over-israel-on-gaza-at-a-low-1407979365. (Hereafter: Entous, "Gaza Crisis").

55 Jen Psaki, Spokesperson, Daily Press Briefing, Washington, DC, July 28, 2014, http://www.state.gov/r/pa/prs/dpb/2014/07/229855.htm

56 Every Day is Extra p. 686

57 Holland Steve, "U.S. Officials Defend Kerry from Israeli Criticism," *Reuters*, July 29, 2014, http://www.reuters.com/article/us-mideast-crisis-obama-idUSKBN0FX1WY20140728

58 David Lerman, "Rice Stands By Kerry After Israeli Critics Attack Him," *Bloomberg*, July 29, 2014, https://www.bloomberg.com/news/articles/2014-07-28/rice-stands-by-kerry-after-israeli-critics-attack-him

59 White House, "Readout of the President's Call with Prime Minister Netanyahu of Israel," July 27, 2014. The next day Kerry stated that "any process to resolve the crisis in Gaza in a lasting and meaningful way must lead to the disarmament of Hamas and all terrorist groups. And we will work closely with Israel and regional partners and the international community in support of this goal." State Department transcript, "Remarks

at the Rollout of the 2013 Report on International Religious Freedom," Secretary of State John Kerry, Washington, DC, July 28, 2014, https://obamawhitehouse.archives.gov/the-press-office/2014/07/27/ readout-president-s-call-prime-minister-netanyahu-israel

60 Susan E. Rice, National Security Advisor, Remarks to the National Jewish Leaders' Assembly, July 29, 2014, https://obamawhitehouse.archives.gov/the-press-office/2014/07/29/ remarks-national-security-advisor-susan-e-rice-national-jewish-leaders-a

61 John Kerry remarks with Ukrainian foreign minister Pavlo Klimkin, July 29, 2014, https://2009-2017.state.gov/secretary/remarks/2014/07/229905.htm

62 White House, Press Conference by the President, August 1, 2014, http://www.whitehouse.gov/the-press-office/2014/08/01/press-conference-president.

63 Rick Gladstone, "U.S. Advises Americans to Put Off Travel to Israel," *The New York Times*, July 21, 2014, https://www.nytimes.com/2014/07/22/world/middleeast/us-advises-americans-to-put-off-travel-to-israel.html

64 Associated Press, "United, Delta cancel flights to Israel; plane in the air is diverted to Paris," July 22, 2014, https://www.mercurynews.com/2014/07/22/united-delta-cancel-flights-to-israel-plane-in-the-air-is-diverted-to-paris/

65 Ashley Halsey III and Mark Berman, "Citing Dangers, FAA Bans U.S. Carriers from Tel Aviv," *Washington Post*, July 23, 2014, http://www.highbeam.com/doc/1P2-36995444.html (Hereafter Halsey and Berman, 2014).

66 Niv Elis and Lahav Harkov, "El Al says there is 'no chance' it will cancel flights to, from Israel," *The Jerusalem Post*, July 22, 2014, https://www.jpost.com/Operation-Protective-Edge/Report-US-considering-grounding-flights-to-Israel-amid-rocket-threat-368492

67 Ibid.

68 Hereafter Halsey and Berman, 2014 *See also:* Anshel Pfeffer *wrote in the regard:* "Whether or not flights in and out of Israel are suspended for any length of time, the suspension of flights by several major air carriers is Hamas' first major achievement of this conflict." Anshel Pfeffer, Will the Threat to Israel's Only International Airport Be a Game-changer? July 22, 2014, Haaretz, https://www.haaretz.com/.premium-will-ben-gurion-threat-be-a-game-changer-1.5256412

69 Prime Minister's Office, July 23, 2014, https://www.gov.il/he/Departments/news/eventblumberg230714

70 Entous, "Gaza Crisis."

71 Jeffrey Sparshott, "Obama Signs Law Providing $225 Million for Israel's Iron Dome," *Wall Street Journal*, August 4, 2014, http://online.wsj.com/articles/obama-signs-law-providing-225-million-for-israels-iron-dome-1407193842

72   Marie Harf, Daily Press Briefing, August 14, 2014,
     http://www.state.gov/r/pa/prs/dpb/2014/08/230614.htm#I.

73   Barak Ravid, "Israeli Officials: Talks with U.S. to Resolve Crisis in
     Weapons Supply," *Haaretz*, August 14, 2014,
     https://www.haaretz.com/white-house-ordered-halt-to-missile-
     transfer-requested-by-israel-1.5259366

74   Marnie Hunter, Katia Hetter and Chelsea J. Carter, CNN US, "European
     airlines suspend flights to Israel," *CNN*, July 23, 2014,
     http://edition.cnn.com/2014/07/22/travel/israel-flights-
     suspended/ (Hereafter: Hunter, Hetter, and Carter, "European airlines
     suspend flights to Israel").

75   Ibid.

76   Jad Mouawad, "Airlines Suspend Flights to Israel After Hamas Rocket Falls
     Near Main Airport," *The New York Times*, July 22, 2014,
     https://www.nytimes.com/2014/07/23/world/middleeast/faa-halts-us-
     flights-to-israel.html

77   Hunter, Hetter, and Carter, "European airlines suspend flights to Israel."

78   PM Netanyahu meets with British Foreign Secretary Hammond, Israel
     Ministry of Foreign Affairs,
     http://mfa.gov.il/MFA/PressRoom/2014/Pages/PM-Netanyahu-meets-
     with-British-Foreign-Secretary-Hammond-24-Jul-2014.aspx

79   *Gaza Blog*, "Kerry tells Netanyahu flight ban solely for safety," July, 23,
     2014,
     http://live.aljazeera.com/Event/Gaza_Blog/122340614

80   Katia Hetter, Pamela Brown, and Adam Levine, CNN, "U.S. ends ban on
     flights to Israel," *CNN*, July 24, 2014,
     http://edition.cnn.com/2014/07/24/travel/israel-flights-suspended/

81   Secretary of State John Kerry Remarks June 15, 2014,
     https://2009-2017.state.gov/secretary/remarks/2014/06/227599.htm

82   Secretary of State John Kerry Remarks July 2, 2014,
     https://2009-2017.state.gov/secretary/remarks/2014/07/228699.htm

83   Ibid.

84   White House Press Briefing July 8, 2014,
     https://obamawhitehouse.archives.gov/photos-and-video/video/
     2014/07/08/press-briefing#transcript

85   Ibid.

86   White House Press Briefing July 14, 2014,
     https://obamawhitehouse.archives.gov/photos-and-video/video/
     2014/07/14/press-briefing#transcript (See also: White House Press
     Briefing July 15, 2014,
     https://obamawhitehouse.archives.gov/photos-and-video/video/
     2014/07/15/press-briefing#transcript)

87   Secretary of State John Kerry Remarks July 15, 2014,
     https://2009-2017.state.gov/secretary/remarks/2014/07/229275.htm

88   President Barack Obama, Statement, July 18,
     https://obamawhitehouse.archives.gov/photos-and-video/
     video/ 2014/07/18/president-makes-statement-ukraine#transcript

89 Ibid.
90 President Barack Obama, July 21, 2014,
https://obamawhitehouse.archives.gov/photos-and-video/video/
2014/07/21/president-gives-update-ukraine-and-gaza#transcript
91 President Barack Obama's Phone Call with Prime Minister Netanyahu,
July 20, 2014,
https://obamawhitehouse.archives.gov/the-press-office/2014/07/20/
readout-president-s-call-prime-minister-netanyahu-israel
92 Secretary of State John Kerry, Interview, *ABC News*,
https://2009-2017.state.gov/secretary/remarks/2014/07/229506.htm
93 Secretary of State John Kerry, Interview, *Fox News*,
https://2009-2017.state.gov/secretary/remarks/2014/07/229505.htm
94 Ibid.
95 Secretary of State John Kerry, Interview, *CNN*,
https://2009-2017.state.gov/secretary/remarks/2014/07/229508.htm
96 White House Press Briefing, July 21, 2014,
https://obamawhitehouse.archives.gov/photos-and-video/video/
2014/07/21/press-briefing#transcript
97 Secretary of State John Kerry, Remarks, Cairo, July 21, 2014.
https://2009-2017.state.gov/secretary/remarks/2014/07/229574.htm
98 Secretary of State John Kerry, Remarks, Ramallah, July 23, 2014,
https://2009-2017.state.gov/secretary/remarks/2014/07/229672.htm
99 Ibid.
100 Secretary of State John Kerry, Remarks, Cairo July 25, 2014,
https://2009-2017.state.gov/secretary/remarks/2014/07/229803.htm
101 Ibid.
102 Ibid.
103 Goodman, Israel's Narrative, 2014, p. 7.
104 Israel Ministry of Foreign Affairs, May 2015, p. 3.
105 Secretary of State John Kerry, Remarks, Paris, July 26, 2014,
https://2009-2017.state.gov/secretary/remarks/2014/07/229811.htm
106 Ibid.
107 Barak Ravid, "Kerry's Cease-fire Draft Revealed."
108 PLO And Fatah: "Kerry Initiative, Paris Conference – A Plot To Bypass
Egypt And Remove The PLO."
109 President Barack Obama's Phone Call with Prime Minister Netanyahu,
July 27, 2014,
https://obamawhitehouse.archives.gov/the-press-office/2014/07/27/
readout-president-s-call-prime-minister-netanyahu-israel
110 White House Press Briefing, July 29, 2014,
https://obamawhitehouse.archives.gov/photos-and-video/video/
2014/07/29/press-briefing#transcript
111 Ibid.
112 Ibid.
113 White House Press Briefing July 31, 2014,
https://obamawhitehouse.archives.gov/photos-and-video/video/
2014/07/31/press-briefing#transcript

114 White House Press Secretary, For Immediate Release, July 31, 2014, https://obamawhitehouse.archives.gov/the-press-office/2014/07/31/statement-press-secretary-humanitarian-ceasefire-announcement-gaza

115 Secretary of State John Kerry, Remarks, Washington DC, August 1, 2014, https://2009-2017.state.gov/secretary/remarks/2014/08/230114.htm

116 White House Press Briefing, August 4, 2014, https://obamawhitehouse.archives.gov/photos-and-video/video/2014/08/04/press-briefing#transcript

117 Secretary of State John Kerry, Interview, *BBC*, August 5, 2014, https://2009-2017.state.gov/secretary/remarks/2014/08/230292.htm

118 President Barack Obama, Press conference, August 6, 2014, https://obamawhitehouse.archives.gov/photos-and-video/video/2014/08/06/president-obama-holds-press-conference-us-african-leaders-summit#transcript

119 Ibid.

120 Ibid.

121 Ibid.

122 Ibid.

123 President Barack Obama, phone call with King Abdullah II, August 8, 2014, https://obamawhitehouse.archives.gov/the-press-office/2014/08/08/readout-president-s-call-his-majesty-king-abdullah-ii-hashemite-kingdom- e

124 Secretary of State John Kerry, Remarks, August 26, 2014, https://2009-2017.state.gov/secretary/remarks/2014/08/230910.htm

125 President Barack Obama, Speech at United Nations General Assembly, September 24, 2014, https://obamawhitehouse.archives.gov/photos-and-video/video/2014/09/24/president-obama-addresses-united-nations-general-assembly#transcript

126 Ibid.

127 President Barack Obama's meeting with Prime Minister Netanyahu October 1, 2014, https://obamawhitehouse.archives.gov/photos-and-video/video/2014/10/01/president-obama-meets-prime-minister-israel#transcript

128 Ibid.

Conclusion

1 Israel Ministry of Foreign Affairs, p. 28.

2 On July 15, 2014, Kerry stated that the indiscriminate firing by Hamas at Israeli civilian targets was against the laws of war and that therefore, Hamas was a terrorist organization. Kerry remarks, July 15, 2014, https://2009-2017.state.gov/secretary/remarks/2014/07/229275.htm

3 Press Briefing by Press Secretary Josh Earnest, July 15, 2014, http://www.whitehouse.gov/the-press-office/2014/07/15/press-briefing-press-secretary-josh-earnest-7152014

4   Interview with Zeinab Badawi of BBC's HARD Talk, John Kerry, August
     5, 2014,
     http://www.state.gov/secretary/remarks/2014/08/230292.htm
5   Friedman, "Obama on the World." *The New York Times*, August 8, 2014,
     https://www.nytimes.com/2014/08/09/opinion/president-obama-thomas-
     l-friedman-iraq-and-world-affairs.html
6   "Transcript: Obama's Full Interview with NPR," National Public Radio,
     June 1, 2009,
     http://www.npr.org/templates/story/story.php?storyId=104806528.
7   The White House, Office of the Press Secretary, "Remarks by the
     President at Cairo University, 6-04-09," June 4, 2009,
     https://www.whitehouse.gov/the-press-office/remarkspresident-cairo-
     university-6-04-09.11
8   Entous, "Gaza Crisis."

Index